MARDIS

MARDIS

THE MANIPULATION INTERPRETATION

T. A. DEAN

Copyright © 2022 T. A. Dean

All rights reserved. This book may not be reproduced in any form, in whole or in part (beyond the copying permitted by US Copyright Law, Section 107, "fair use" in teaching or research, Section 108, certain library copying, or in published media by reviewers in limited excerpts), without written permission from the author.

Printed in the United States of America

Cover and Interior Design: Creative Publishing Book Design

ISBN Paperback: 979-8-9856299-0-3
ISBN eBook: 979-8-9856299-1-0

Disclaimer

This book is a memoir. It reflects the author's present recollections of experiences over time. Some names and characteristics have been changed, some events have been compressed, and some dialogue has been recreated.

This work depicts actual events in the life of the author as truthfully as recollection permits and/or can be verified by research. Occasionally, dialogue consistent with the character or nature of the person speaking has been supplemented. All persons within are actual individuals; there are no composite characters. The names of some individuals have been changed to respect their privacy.

To my family.
Bird & Lulu
&
every adventure that remains.

Acknowledgments

Special Thanks:
Tammy Kanne, Chris Moore, David Levin, Nikki Moustaki, Brittany Brady, Tanisha Washington, Thendral Kamal, Rehanna Lehman, Robert Bearden, Nadine McKinnon, Tina Daste, Christine Ferentz, Shannon Bistline, Claude & Janice Davis: the words, thank you will never be adequate to share how much I appreciate you or explain how much you have meant to my spirit.

Special recognition to:
Chai Camillo, TJ Mumbower, Shelley Defibaugh, Mark Matthews, Vickie Thiel, Susan Conditt, Linda Chrisman, Julie Williams, Regina Isaac, Larry Wooters, Steve Greenaway, Dan Skaggs, Bob Alan, Robert Whitkop, Orchid Bogaert, and Emma Flax.

With additional thanks for editorial assistance:
Ghislain Viau, Joni Wilson, Vincent Simonette, Keidi Keating & Sarah

Contents

Introduction.................................... 1

Chapter 1: Redemption 9

Chapter 2: Anxiety............................ 23

Chapter 3: Faith.............................. 41

Chapter 4: Connections........................ 49

Chapter 5: Hypocrisy.......................... 69

Chapter 6: Answers 81

Chapter 7: Isolation........................... 93

Chapter 8: Charity........................... 101

Chapter 9: Manipulation 113

Chapter 10: Restoration....................... 129

If You Need Help............................. 143

Introduction

It doesn't matter who we are; there's always someone in life that will cause anxiety. Whether we are struggling financially or are wealthy makes little difference, life can feel hopeless to anyone. Human beings have a silent instinct, and we know that life doesn't have to be a continuous battle against stress and depression.

Deep down, we know that finding happiness shouldn't feel impossible or so far out of reach. Yet, daily we can struggle, and many have dark feelings of despair. However, love and compassion restore balance to our world, and those qualities can effectively numb emotional pain.

One of the many challenges we humans face in finding and achieving our balance is that we often forget that we're in control. We choose what we share and how we behave and react. Subtle influences around us encourage us to let go of that control, but the choice to let go is ours to make and ours alone.

We fail when we refuse to take the necessary time to consider our actions and reactions carefully. Our vanity also encourages us to search for a villain to blame when our choices lead to negativity. The stress that tends to follow promotes other terrible decisions. Face it; there is no hiding negative behavior. Others feel the impact of our terrible choices. We perpetuate anxiety in a seemingly endless and vicious cycle as a species.

Yet, some people will intentionally live a negative existence. Many of these people desire and crave the creation of stress and actively cause mayhem in the lives of others. Somehow their negative behavior brings comfort to their spirit. Those behaving this way are usually those with the most influence in our lives, whether they deserve that power or not.

Many of these people are also somewhat convincing. They'll use our impatient nature, lack of context, ignorance, and desire for quick answers to alter our perception. They play in the gray areas of our uncertainty, and we don't realize that we have been manipulated or have turned over control until it is too late, and the consequences of our actions begin to sting and burn.

We slowly begin to believe that we don't matter, but this belief can only form when we purposely embrace the idea that despair is all we deserve. We do count, and we are worth more than we might realize. In time, as we evolve, even the most impatient among us will recognize and understand the shame that ultimately comes from our poor personal choices and our failure to act with honor.

• INTRODUCTION •

When the person who causes us the most stress in our lives is a relative, it's hard to disconnect, move on, or grow. Whether we like it or not, family is tied to us forever. Ethics, morality, and values are not always consistently shared in families. Being related is never a guarantee of peace or mental stability.

For a long time, I felt outnumbered. I lived in a constant fight against ignorant ideas and insanity. I was surrounded by those who savored the delicious emotional harm they would cause others.

As a child, those closest to me were religious extremists, bigots, pedophiles, and apathetic people pretending to be stable members of society. I was unaware of my internal authority; I became bitter. Deep down, I knew something was wrong and didn't want to behave the same way as those around me.

Growing up, I felt weak and drained from constant mental attacks and unending pettiness. I eventually reached a point of depression where I genuinely wanted nothing more from life than to end it. I never embraced the awful behavior I witnessed or even pretended to approve of the constant insanity. I was angry because I watched pain unfold daily and felt powerless to do anything about it. By thinking contrary to those causing pain, I became a target for further and more drastic abuse. Nonconforming always meant harsher treatment and agony.

Eventually, I felt trapped and unworthy of happiness. Yes, I even accepted that I would always fail, as those around me consistently predicted. Even when genuinely caring people came into my life, I would push them away. It's not that I didn't need beautiful and loving people in my life, I did, but I moved them out of my life to spare them from my existence. I was sure they deserved better than me. I was determined to protect those caring people from the infectious madness of my isolated world.

I stopped allowing anyone to get close. I continued to monitor every example of love and compassion from a distance. In silent admiration, I began watching good people from afar. The longer I watched, my withering hope seemed to slow. I became inspired by teachers, hairstylists, bus drivers, colleagues, and even strangers who showed random kindness. People of morality and decency were beacons, and each lit my way. I clung to each memory and every act of considerate behavior, but my journey was a slow process of wandering in the darkness of hell, alone.

A significant turning point in my life came when my well-meaning sister convinced me to visit one of the meanest people we knew, our dying grandmother. She was a person who never showed either of us an ounce of compassion. She was emotionally abusive and was all too often without reason.

While I admit she might have simply suffered from a loss that I was never aware of, her manner of dealing with her troubles became taking them out on others. Eventually, I held her responsible for many of my unnecessary anxieties as a child. She became my nemesis, and our toxic interaction gave me an excuse for my poor behavior and awful personal choices as an adult.

When I made mistakes in life, it was always easier to use reasons from my past and blame the villain. Instead, I should have taken responsibility for my many failures, but I became more like those who caused me pain than I cared to admit. At one point, I thought it was too late and that I would never be able to find my way out of the darkness.

As my grandmother's life was coming to an end, I needed to decide if I would forgo the opportunity to say goodbye humbly and with decency or if I would be apathetic and say good riddance. I made a choice, but I didn't know what to expect. Something good must have still existed within me, and I chose compassion. I admit my reluctance almost won the day, but the visit was worth that internal fight.

My grandmother was certainly not the sweet variety. She never baked cookies for her grandchildren, knitted sweaters, or sent birthday cards with hugs and kisses. She had an eternal rage. She was also a meddler and gossip. As long as we knew her, she treated my sister and me like we were worthless.

Although I can't speak for my sister, she treated me as if I were the spawn of Satan and the most unwanted person who ever dared to breathe and was unworthy of God's air. Her name was Mardis, and I didn't think that we could ever call a truce or learn to trust each other, but that was about to change in an unexpected, deep, and meaningful way.

Revisiting Mardis was not only the right thing to do. It was incredibly enlightening. In her final days, she managed to teach me some of the greatest lessons I might ever learn. Within hours, I developed a genuine love for her, and I didn't want to leave her side. I reached a point where I would miss her sincerely when she finally passed away.

It took nearly a lifetime to reach this point of love, and I was now devastated that our time was up. That final week changed me, and I hope to share our story and some of what I learned. Healing is indeed possible, but each of us shares responsibility.

Closer to her end, I felt as if her soul somehow managed to hold open a door where the universe's secrets were hidden. She appeared to be wedged between life and death, stuck between two realms. She revealed an understanding of life that is still impossible to define clearly, but she shared a fresh perspective on practically everything.

Her gifts of wisdom ranged from our emotional demons to the unwritten laws and patterns of the universe. Sadly, our strange and entangled communication was interrupted as her passing finally closed that door, and her strength eventually failed.

• INTRODUCTION •

Our final moments together forced me to reflect on each of the burdens and blame that I carried. I can only assume that Mardis was trying to guide me to a better way of life and happiness before her life ended. This gift might have been a final act of repentance, but the change in her was incredible. She didn't seem to be the same person. It was as if she was from another universe.

The difference she made within me would take a little longer to understand. Still, I could feel a change from her final influences spreading within my soul as the essence of her thoughts slowly began to fill my heart and mind. Those final bittersweet moments caused a harmony that resonated in every atom of my body and strengthened my spirit.

The only way to establish a reliable example of how pain can be overcome is to share and confess my own. Hopefully, these memories of redemption will provide comfort to others. I'm not attempting to discount the pain of anyone who is currently suffering, but I need to show that this text is not just posturing words alone.

A strength could come from such end-of-life revelations, but through personal effort, honesty, and self-reflection, an incredible power also emerges within us. This power is slightly out of phase from our daily understanding, but it's always there. With the help of others, we can get a step closer to seeing it and be more open to enlightenment. If sharing these thoughts can help a single person, maybe it can cause a positive change in the world that is everlasting.

At this moment, you might be facing a similar struggle in your personal story. Please remember, regardless of how isolated you become or how lost you feel, this attitude can never last. Many souls are entangled with you and feel your pain; others will reach out to guide you, but these souls can only help if you allow them to try.

CHAPTER 1

Redemption

Mardis was part of a reasonably large family: three sons and three daughters, each of them have so many children that I can't count them all. My sister and I lived closest to Mardis. Although we spent more time near her, she never said a kind word and would never bother to visit. She was extremely strict and unforgiving, and her name was hard on the ears. For decades, her name gave us anxiety. When it came to our relationships with her, we always asked, "What horrible thing has she said about us now?"

Whenever we went to forced gatherings with her, she ensured we would only hear how wonderfully behaved and highly accomplished her other grandchildren were. She marveled at their every story, contrived or not. If we attempted to share anything we were proud of, she would immediately interrupt us and change the subject to the achievements of others before we could finish our sentence.

MARDIS

We always knew we would never hear such praise and got used to being last and treated the worst. Our mother was the eldest child of our grandmother and held a special place in her heart, so having grandchildren who didn't measure up to her idea of perfection was a disappointment.

My sister and I were frequently left alone by our mother for many days at a time. Our estranged mother would leave a twenty-dollar bill on the counter and disappear, often flying away to unknown locations with a boyfriend, and we never knew when she would return. She was having a highly open affair with her boss, a married man. There were no cell phones or fast ways to reach one another during that time. We sat by the phone and waited to see if she would call, but the phone never rang.

My older sister was barely ten, but she was in charge of our care when the affair began. We never knew when our mother would return, so we got used to being alone for days, and these absences went on for years. While Mardis lived a short drive from us, she never checked in on us, and we became latchkey kids—children without adult supervision.

As our mother's affair continued, my sister and I sadly learned that her boyfriend was also a pedophile. Our mother turned a blind eye to his demented abuse—there was too much at stake. She knew losing him would also mean losing the lifestyle she'd become accustomed to and felt she deserved. There was no way leaving this man was a possibility, so the sexual abuse we endured continued until we attempted to reach out to the only other adult in our lives: our grandmother, Mardis.

Speaking about our sexual abuse out loud was a mistake. Eventually, our concern found its way into the real world. While Mardis wanted us to keep it hidden to protect her daughter, state and child protective services got involved somewhere along the way. The indictment was sexual abuse of a minor, but only for my sister's abuse. No one bothered to ask how extensive the abuse was or considered that this miscreant was equally abusing boys and girls. It was so embarrassing and damaging. I lost trust in many people, and I remained isolated for decades. I was frequently called antisocial, but the reality was, I was struggling and in deep emotional pain, but I felt unworthy of anything more than misery.

Before their day in court, our grandmother convinced my sister to lie about the abuse. By that time, my sister was fourteen. Several years of abuse occurred, yet Mardis asked my sister to lie in court to protect a raging pedophile. Our mother's income and freedom depended on it. The exchange between Mardis and my sister was the first time she ever spoke to either of us in a way that could be considered polite or kind. This single manipulation led to the abuse of many other children.

As children, there was nothing we could do. Our life paths altered forever, and these events might have kept us from our purposes in life. Every action where a selfish desire seems narrowly consequential grows exponentially, permanently altering and destroying the lives of many. Some consequences span generations in the future because of a single act of selfishness in the past.

In the end, my sister did as she was told and lied. She testified that she was merely looking for attention. She covered up her abuse and never brought up mine. I couldn't discuss my abuse with anyone because all the focus was on my sister. I was ashamed and terrified and would not have been brave enough to speak out anyway at that age.

Mardis seemed to be protecting her daughter to avoid child endangerment charges for allowing sexual abuse. She was angry we even brought it up, and she was bitter over the entire situation. Our need for protection seemed like attempts to make our mother's life miserable and nothing more.

Mardis was the only relative in the same city and state, but this scandal was unsavory, so none of our family would ever hear about it. If there were a chance that the information would make our mother look bad, the story would never see the light of day. If our mother went to prison, responsibility and custody would fall to Mardis, which she didn't desire in the least.

The case was over quickly, and the lawyers patted the monster on the back as if they were best pals, smiling and shaking hands as the doors of the courtroom closed behind the victims. Our mother was also spared and didn't receive any criminal sentencing. My sister was required to undergo mandatory counseling. The court claimed she needed counseling for a "lying disorder" and need for attention, but we knew better. Everyone just silently turned and walked away, and the case was closed.

The abuse stopped for my sister and me because we were too much of a risk, but the abuser, our mother's boyfriend, continued with scores of fresh new victims, as we learned later in life.

Because our mother worked for her boyfriend, she kept her job in the glamorous world of trailer park management, where she fancied herself a chief executive officer, collecting rent checks from tenants. She continued to have the lifestyle she wanted and a job requiring no special skills. Her position only needed a calculator and receipt book, but she was a titan of industry in her mind.

My sister and I were now considered absolute garbage in the eyes of Mardis and our mother. We were nothing but trouble for mentioning scandalous sexual abuse out loud, and it was all our fault. It was not until much later in life that we learned why our mother was less than a stellar parent. We discovered that she was bipolar, was learning disabled, and was a malignant narcissist, and her condition also included pathological lying. She was essentially mentally incapable of understanding love or empathy for others.

Mardis was once a nurse and might have recognized the illness. As her mother, Mardis might have known about the diagnosis and hidden it from everyone, which could be why Mardis chose to stand by her side. Staying close to her daughter would offer protection from the next awful mistake she was bound to make.

Only Mardis ever really knew why she remained so close, but there was a shared psychosis, a madness they both seemed to enjoy. Their apathy seemed infectious to others and inspired awful behavior in many, spreading far and wide. The scheming duo were experts in the art of manipulation and relished causing despair in the lives of others.

Mardis spent each day with our mother at the trailer park, and this gave them both a buffet of souls to torment. They would frequently insert themselves into the problems of those living under their watchful eyes, and they relished their perceived power to evict anyone who got out of line.

I initially saw this behavior as a usual way of life as a child. The only decent examples of parenting or morality available to me were in the context of carefully scripted television shows that aired lessons on the differences between right and wrong. Actors on television portrayed parental roles, but they were parents by proxy, and I needed them.

• REDEMPTION •

Other examples of morality came from the parents of some of my few friends. I was very confused when I began to notice the basic moral decency of other families. I struggled with those parenting styles and the nurturing support I witnessed, but I knew neither my sister nor I would ever know such nurturing or sincere love.

As we became adults, we still endured verbal abuse, but this was significantly less bothersome due to our adult responsibilities of work and raising our children. Although we no longer lived in hell, we continued to hear the occasional rumors about how terrible our lives must be. Still, the context began to evolve to cover our reluctance to visit them.

As a result of our absence, our mother and grandmother would take turns proposing new stories about our failures, almost wishing despair upon us. They often discussed what we must be doing, making up our potentially lousy behavior, with each overheard tale far worse than the last. Their devious gossip continued until something both tragic and wonderful happened.

The tragedy was when Mardis developed dementia. She was living with our mother during this period. Age took a heavy toll on her, and she was incredibly frail. Despite Mardis's rapidly declining health, our mother took advantage of Mardis's weakened mind and cruelly forced her to become a constant servant.

• MARDIS •

Our mother would demand that Mardis prepare food, bring beverages, clean the house, and do her laundry. Before Mardis was required to move to the hospital, my sister visited. At one of her most honest and transparent mental moments, and before dementia destroyed her thoughts, Mardis looked at my sister and weakly confessed, "There is just no pleasing her." Meanwhile, our mother continued to maniacally bark orders from another room.

My sister stopped Mardis from working, took over her so-called chores, and laid her down to rest. But the following day, Mardis was in an ambulance and rushed to a hospital when she collapsed from exhaustion and severe dehydration. Her condition was much worse than our mother cared enough to reveal.

My sister was the first to visit Mardis in the hospital, and, for a day or two, she was her only visitor. I was still distant and in denial. I didn't believe dementia was her problem and thought her condition must have been another elaborate con. Although I trusted my sister, I reluctantly agreed to visit Mardis and prepared to be fully mentally guarded.

During my sister's earlier visits, she and our grandmother truly bonded. There was finally a sense of genuine love, and our grandmother could see my sister for the kind and nurturing person she always was, but time was running out quickly. Still, there was a lighter feeling around the two.

• REDEMPTION •

When I walked into Mardis's hospital room, I could hardly believe that this was the same person I always feared. Her skin was mushroom gray, and her hair looked as though she tried to cut it herself—crooked and thinning. She was emaciated, but she was also so messy and untidy that I could barely recognize the woman.

The enormous shock of seeing a person who was always strong, cold, and abrasive in such a weakened state was devastating. I felt an immediate sense of sadness and pity. I couldn't believe it was her, but it took less than thirty minutes for me to confirm what my sister already instinctively knew: she didn't have long to live.

I was extremely fortunate to arrive when I did. I managed to catch Mardis in one of her few remaining moments of clarity. We found forgiveness for each other during this rare period, and she held my hand. I didn't know what to do. It felt awkward at first, but later, it was as if we were speaking from one soul to the next while exchanging unspoken memories and thoughts. Mardis shared the pain from a life filled with regret, and I felt each moment.

I suddenly remembered some breakfast cookies in my pocket. I shared them with Mardis; she seemed to enjoy them. As she looked at me with crumbs on her face, she spoke one of her last clear thoughts: "The burdens of life are heavy and are too much for you to carry. You've carried this burden too long.

Just let go. It's time to move on and rediscover your purpose." She continued to hold my hand even though it was sweaty and felt awkward. I couldn't move. I just sat silently and still and wondered what would happen next. I debated if I should try to move my hand, but she needed me no matter how uncomfortable I was, and she continued to share her thoughts.

While my grandmother's behavior in the past might have been contrary to her Bible teachings, she once knew the text well. She kept a tiny plastic loaf of bread on her kitchen table when I was a child. The plastic bread contained small cards with various biblical Psalms and Proverbs printed on them; she called it the "Bread of Life." I remembered this, so I managed to keep her busy, reminding her of some of her favorite scriptures. When my sister saw this exchange between us, she was amazed that the two of us knew so many similar things. To think that Mardis and I were anything alike was bizarre.

At the end of each workday, I returned to visit, and I'd stay for hours until she would fall asleep. Neither my sister nor I wanted Mardis to be alone. We saw her in turns; no one else could be bothered.

On one of my visits, the nurse announced, "Your grandson is here." When Mardis saw me, she scoffed and said, "No, that's not my grandson. That's my minister." She genuinely didn't know who I was any longer, but she thought enough about me to see me as a person of peace, which I needed badly.

• REDEMPTION •

It didn't matter that she didn't know who I was. All that mattered was that we found forgiveness together while her thoughts and heart could allow it. I can honestly say I loved her at this point, deeply. I was happy to be with her until her end.

On the night of her passing, I dreamed that she might be saying, "Goodbye." I was driving down a highway, approaching a long bridge over water, but leaves were falling all over the road. There were no trees around. The radio played the song "Carry on Wayward Son" in my dream by Kansas. I remember thinking Mardis would never approve of this, as she would have once called it "dirty hippie music." Yet, the moment I began to think this in the dream, I heard her whisper a long "Shhhhhhhhh," and it felt like her presence was slowly fading away.

To this day, when I see a leaf falling, regardless of where I am, I feel that it's her, and I now feel warmth and love in my heart. When I listen to the song's lyrics, I understand why she would have chosen it to say goodbye. The song left guidance that she didn't have time to share before reaching the end of her life. I am grateful for what might have been a loving farewell, and I am confident she made it to her place of peace.

While healing often seems impossible, it does happen. Despite our pain and circumstances, each of us found some forgiveness. We were fortunate to share in this redemption;

the hate and blame are gone forever. My visits with Mardis also reminded me that we have no choice but to tolerate our current existence, regardless of circumstances, and we should make every moment count.

I firmly believe we have a moral responsibility to lift one another as we continue our journeys through life to its end. As time runs out, each of us will eventually reach a final destination just like Mardis, so we should never squander the time we have. Blame takes entirely too much energy from us, and we are better off using our power and energy for healing. Suffering ends for everyone, but no longer will we endure—as William Shakespeare put it—"The heartache, and the thousand natural shocks That flesh is heir to" (*Hamlet,* Act 3, Scene 1).

As our energy transforms and we ascend to that same plane of existence of those we have lost, broken hearts will mend. Our enlightened spirits always feel compelled to be with others in their darkest moments, transform suffering, and nurture our souls. This moral exchange is love, a gift, and we give it freely.

Although we might not immediately realize when our souls entangle in this way, we instinctively act as messengers of hope each time we are needed. Honestly, I wasn't there for Mardis; instead, I believe she chose to be there for me.

• REDEMPTION •

Were my final encounters with Mardis and forgiveness a series of miracles? I can't say. Was Mardis saying goodbye? I can't be sure of that either, but I needed to share this pain in this book because it might help many seeking answers now. Perhaps our story was also told to help you on this exact day.

CHAPTER 2

Anxiety

Some might feel strongly about how Mardis treated my sister and me as children. Many might think we should consider her end as karma, but we know this is not true with enough time to reflect. Our anger encourages us to say so many things we don't often mean.

Our unbalanced understanding of most situations confuses us, and we often begin to feel and behave contrary to our better selves. We know we are better than our worst traits, but we'll always have to fight that good fight within. This internal battle will continue until our lives are finally over, but the fight gets easier with every passing day.

Before our time with Mardis, we were a transient and homeless family. Our primary battle was avoiding starvation. We slept in an old bread delivery van and spent most of our time on the road, traveling from state to state. We temporarily stayed with relatives from our father's side of the family and frequently visited his old war buddies from his time in the military. We would move again when we overstayed our welcome; this cycle continued for years.

Trash dumps were our shopping malls, and the swamps, bayous, and oceans became our grocery stores. Occasionally we chose to live in tents with many creepy crawlies. These critters visited us daily, which caused a lot of tension between our mother and father. Our mother would scream dramatically about anything natural, but our dad didn't mind these critters at all.

I can recall countless nights of our parents yelling for various reasons. At night, our mother would scream about the snakes and spiders, and our dad, a veteran of Vietnam, endured fits from posttraumatic stress disorder (PTSD), where he often woke shouting with flashbacks from the war. The two were at each other's throats every single day, constantly arguing. There was no such thing as sleeping peacefully or having sweet dreams. Waking suddenly to loud shouting was terrifying and torturous. This type of life and sleep deprivation settled my mind in a place of chronic and endless anxiety.

My parents were both having a rough time. Their lives were nothing like they dreamed they could be or should have been. Our dad trained to take orders as a soldier, and our mother loved giving them, but neither could cope with how hope in their lives vanished; life became a long series of disappointments.

Our parents' disappointments were also ours as children. We never knew when we would be too cold or too hot, what we might eat, or where we could go to the bathroom. To this day, I'm terrified of snakes and spiders. I also thoroughly dislike fish, and the smell of seafood makes me ill.

The natural nightly visitors seemed drawn to me. Many nights, as we lived a tent life, wildlife would curl up with me or crawl over me, and our mother would become violent at the sight of them, screaming in terror.

Our father kept us on the move as if he were still on tour in Vietnam, and we lived off the land for years until I was closer to school age. My poor sister simply didn't attend school most of the time when our mom and dad were together and on the road. They eventually called it quits, and that is when we met Mardis. Although I always suffered from high anxiety, when we met Mardis, it grew.

While we no longer lived with snakes and spiders, our new fears became the darkness, demons, and the devil. When Mardis heard about the snakes and their attraction to me, she believed the serpents chose me to do their bidding. She was immediately suspicious and hypercritical of anything I said or did, and she saw the snakes as a sign of dark demonic forces at work and instantly made me feel like I was evil.

I was never afraid of the dark before, but anything unseen or in the night became terrifying to me. The longer we stayed with Mardis, the stronger those fears grew. I couldn't walk by a dark bedroom without getting the jitters, and I always felt like someone was watching me and waiting to strike me down or drag me off to hell because I somehow belonged to the devil, and the snakes offered absolute proof to Mardis.

Mardis was highly religious, but she disliked us, and she preferred living alone. If we planned to stay with her, she felt it would be necessary to introduce us to her faith; she was a "Southern Baptist." Her brand of belief was filled with constant threats of death and a guarantee of burning in hell's fire for eternity.

If we ever did anything that she didn't like, her temper was quick. Although she wasn't physically abusive, she easily scared us into submission. She would frequently pray over us while pressing hard on our foreheads and loudly ask God to release us from the demons responsible for our behavior. She would continue these rituals until we acted appropriately or until she simply got tired. It's surprising how some people would see this manner of conduct as necessary to please God.

Mardis possessed a few gentle moments, but she was highly vocal about our presence, especially the financial and personal burdens we created. We almost felt like orphans and were threatened with the idea of living in an orphanage many times. Mardis and our mother frequently made us pack small suitcases and drive us away. They pretended to take us to a drop-off point to be picked up by others. We believed each day might be our last together. Every time we packed was like a practice run for the real thing.

Discipline was always followed by shouting and frightening prophecies of our small frames and flesh burning and peeling away. I was barely in the first grade when introduced to what Mardis called faith. When I became an adult, the anxiety from the promise of searing flames clung to me like roots growing through my spine. I encountered many worries, but faith was one of the strongest and most frightening.

My morning ritual as an adult was to feed a horrible little demon—anxiety. I'd spend each morning negotiating with my fears and doubts about everything. Stress consumed every hope, all of my time, and any happiness I dared to feel. My pressures demanded I devote myself to constant worry. Because anxiety is hyperbolic, fearful ideas are always inflated to extreme levels. Even the slightest curious or random thought had the potential to cause a large amount of fear.

Later in life, my studies in anthropology helped me understand anxiety and how it was once beneficial to humans. It once helped our species develop apprehension about dangerous situations. Humans worry less today about predators, and while life is not perfect, we have access to a few of our basic needs. As a

• MARDIS •

result of progress, anxiety got demoted from its once-dominant position in our lives. The evolutionary demotion forced anxiety into concerns far less detrimental to our survival, so it exaggerates any minor problems we have to remain relevant.

My life was transitioning with the introduction of Mardis. We went from a homeless family to the extremism of her faith, and this change seemed to grant my anxiety a kind of perpetual authority over my mind. I lived with it for many years until I learned how to fight back, but that was a lesson that nearly killed me.

Several years after Mardis passed away, I developed a tumor the size of a tangerine in my colon. The tumor tore open my intestines, and sepsis followed, but initially, I only felt horrible flu symptoms. After a few days, I was violently shaking and shivering uncontrollably. I couldn't eat and didn't want to get out of bed. I fell in and out of consciousness for a couple of days; I was in absolute agony.

My skin was turning the same shade of gray that Mardis had before she passed away. I stressed over calling out sick for work, but I still didn't consider how seriously ill I was. I couldn't do anything or focus while in pain, and my confusion became so pronounced that I knew something more must be happening. During stressful times, anxiety typically visits and fills my head with those bit of exaggerated fears, but this time was drastically different.

• ANXIETY •

As the pain became too much for me to handle, I finally relented, and my wife drove me to the emergency room. The hospital administered a computerized tomography (CT) scan and confirmed that I was in worse shape than I thought. I was too stubborn to go when my wife suggested it days ago. I waited too long to see a doctor, and now I might not survive. I wasn't scared. I was relieved and ready to check out of my existence entirely. After a lifelong emotional struggle with anxiety, nothing seemed to matter other than leaving everything behind so I could rest in peace.

I thought:

"This could be my only way out of what seems like a never-ending and utterly stupid fight."

Happiness always seemed out of reach anyway, so if I were to die, my illness would finally end my desperation. Because I was always obsessed with ensuring my family was taken care of financially, I thought they would be fine without me and better off.

While I resolved a lot of the pain in my life since Mardis, my spirit continued to lack confidence and stability; my personal life always felt tenuous. I was simply ready to leave everything behind and wondered if this was what Mardis must have thought as she reached out to me in her final days.

I was eager and fully prepared to consent to a do not resuscitate (DNR) order. As I made that decision, my inner demon—anxiety—seemed to vanish. I was being prepared for surgery and filling out what seemed like hundreds of forms, and the DNR issue finally came up. I was about to close the deal because I just didn't want to go on, but one curious young nurse fought against this idea and favored the continuation of my life.

She was a confident young woman with short dark hair and emerald eyes, and her voice revealed a gentle way of speaking. We were alone in the room, and I candidly told her my intention and my reasons to sign the DNR. The nurse was determined to change my mind about giving consent. Although this was the first time we met, she acted as if she would suddenly lose someone she deeply cared for, it was strange and seemed somewhat unrealistic, but her argument was touching.

She tried her best to reason with me. I saw a tear run down her cheek as she continued making her case. She then held my hand. It was a bit uncomfortable, but she made her best argument for me to refuse to sign that DNR order. She continued to tell me there were so many reasons to live. She spoke of my wife and said I had much more to accomplish. I was delusional from the pain, but I willingly continued to listen. She urged me to fight the temptation to leave everything behind.

She left me alone for a few moments to consider her arguments, but I decided to follow her advice when I saw her cry. Everything went still, and there was a kind of darkness on the periphery of my eyesight like I was beginning to pass out. It could have been the infection, but I seemed to slip into a long, slow daze, and time seemed to stop completely once the door closed.

Suddenly I heard the door open again, and everything was moving, and the sounds were gradually beginning to catch up and match each movement as a new nurse entered the room. At first, I thought it was that convincing young nurse, so I blurted out my verdict: "OK, I'll try to hang on." The new woman seemed confused by what I said. I explained the earlier DNR conversation, and her confusion seemed to lift. The new nurse gave me the same batch of forms as if we were going over them for the first time.

I never saw the other nurse again. It was impossible to know whom I spoke with just a few moments before the second nurse entered the room. I was highly delusional from the infection and high fever, and I couldn't remember her name. It must have been déjà vu, but the medical red tape was also ridiculous. My sense of time continued to distort. I felt like moments were looping, and I somehow lived these exact events many times before.

After I finally got through the endless forms, the staff rushed me to surgery. There was no longer a choice. I learned from the surgeons that the tumor was likely colon cancer, but I was in immediate danger from the infection. Life, I thought, would be radically different for me after this surgery if I survived it at all. I didn't know if I would wake up tomorrow, but I no longer cared to think about it. If I were going to die, the first person to meet me would likely be Mardis.

I accepted everything happening around me in a blasé manner; it was as if I surrendered myself. I watched the ceiling tiles pass by as hospital staff rolled me down the halls and into the operating room, and my sense of time seemed to change again. Some moments were faster, and some seemed to last much longer than they should. As they pushed me through the surgical room doors, time froze one last time.

I began to hear the surgical team's muffled voices under their masks, and I followed their instructions to breathe deeply. I took one last slow look around the room as someone leaned in and whispered, "Shhhhhhhhh," and I went dark.

When I woke up in recovery, the first thing I noticed was a cold plastic bag attached to the right side of my abdomen. I was not expecting this—the bag filled with a dark black liquid that looked like old motor oil. Typically, I would have felt violated by such a significant change in my body, but I felt nothing. I didn't have the slightest hint of depression, but I maintained a dull and confused frame of mind. It was all like a bad dream.

The doctors visiting after surgery began discussing the need for chemotherapy and confirmed that I did have colon cancer. I was facing more uncertainty than ever before, yet

• ANXIETY •

I was alive for some reason. I kept asking myself, "Why am I still here?" The only emotion I felt, however, was guilt for being alive. I didn't know what else to think, and my mind remained in a surreal state of silence; I was simply numb.

My new situation wasn't the stupid and hyperbolic variety I was accustomed to in the mornings. I was in real trouble this time, but I wasn't a hostage to the petty fears I once felt and fed each morning. Gone were the days when I would wrestle with thoughts like: "The world hates you, you'll lose your job, you'll be homeless again, and you're worthless; no one could ever love you." Each of these thoughts became obsolete.

My life was still in real jeopardy. Although the tumor was removed, I was still fighting off the infection caused when the tumor tore open my intestines, allowing days of filth to saturate my body. Still, the little thoughts I was so accustomed to each morning were silent.

I was once somewhat famous for my views on life and would sarcastically say, "Let's wrap this up!" Only my new situation was no longer a joke. There was much more to think about, and I was extremely weak. I somehow made it through surgery, but there was no bright and calling light in that darkness. I went from extreme silence to a hospital room connected to tubes and wires.

If my existence continued, it would have to change. I didn't want to spend any remaining time in a constant state of worry. The seriousness of my new diagnosis made me see my previous anxieties as awful memories and utterly pointless, but I couldn't stop thinking, "Why am I alive? I shouldn't be alive. I'm not supposed to be here."

During my hospital stay, previous toxins that I once frequently allowed in my body purged. I was on various drips, like saline and antibiotics, and I was not eating solids. I received oxygen, and I finally slept without nightmares. Thanks to those fluids, the usual items that I wrongly believed managed my anxiety, like caffeine and nicotine, were flushed from my system after a few days.

I also refused external stimulation from television, phone calls, and social media. I felt like a blank template of myself; I was still me but not me. Not only was I purging harmful chemical substances, but I was also digitally fasting and cleansing my mind. I cut myself off from everything and every influence. My previous petty anxieties remained silent.

Diet, hormones, chemicals, and daily stresses—or a combination of them—contributed to my failing physical and mental health. All I now focused on was recovery and rest. I spent the time in deep reflection and meditation. The last time I was in a hospital was with Mardis, but in my case, I was slowly beginning to realize that I had a second chance, and I wondered what I would do with that opportunity. I didn't feel like the same person.

• ANXIETY •

Yet each night in my hospital bed, I would still look at the ceiling and say, "Please just let me die." My mind needed time to adjust, heal, and compensate for those missing pieces. Even something as toxic as stress requires an adjustment period when it is gone. No, I didn't miss anxiety, but being so used to it and not having it was surreal. I needed to get accustomed to life without that old anxiety, which was a habit that became a necessity in my life over many years.

After the hospital released me, I was allowed a month of medical leave to adjust to the world again. Although I was prescribed pain medication, I refused to take any. While it hurt like hell to move too quickly, pain management has a tradeoff. Easing my pain from surgery through medicine would have slowed the natural healing process. I needed to heal quickly.

My mind was finally beginning to clear, and I reached a point of meditation and peace for the first time where I could allow myself to feel like I truly wanted to live. My brain chemistry was finally balanced, and I didn't want to ruin my fresh new clarity with pain medications or other influences.

My cancer diagnosis didn't bother me. I simply saw cancer as just another change, but I realized I could live in happiness and be free from the usual fears that haunted me. I predicted my original anxiety might eventually return; I couldn't stay on fluids forever. I made observations of the specific elements that would encourage it and made myself a lab rat. I needed to learn as much as possible about how it happens. I wanted to avoid getting trapped again in a mental hell. I knew anything I put into my body and mind would have an impact.

I don't recommend cancer to manage anxiety, but there were many tangible things I learned, such as how anxiety survives, when it appears, how it operates within me, and how to shut it down.

Stress is a chemically induced reaction. While it is true that certain things we hear, see, and feel might cause stress, the severity is driven chemically. My refusal to reintroduce harmful chemicals and accept unknown impacts from medications on my brain chemistry helped me see and understand that chemical link.

Although my natural chemical state was finally in balance, I knew I would continue to feel nervous and anxious even though the toxins purged. Anxiety continues until we work through the remaining psychological artifacts caused by chemical reactions. Understanding this helps to reduce the duration of stress.

We only prolong and reward our anxiety when using chemicals that disrupt our natural balance. Rewarding those urges always leaves us in a constant need for more, and we only gain a false sense that substances like sugar, caffeine, nicotine, or other items are helping. Even a substance as common

• ANXIETY •

as sugar can severely disrupt our balance in dramatic ways and can change our bodies' rhythms and chemical densities.

But our power and focus increase after those substances are purged from our systems. The massive amount of fluids I received and the self-imposed digital fasting from phones, televisions, and the internet helped that process substantially. I wasn't putting anything in my mind or body, and I used the time to rest and reflect. It helped to reboot my mind to the point of calm naturally.

Avoiding anxiety and stress requires paying close attention to what we eat, drink, and, most important, what we mentally consume. Everything from nutrition to information is essential in maintaining a favorable mental balance. Anything that goes into the mind and body will eventually impact our bodily functions, leading us to an entirely new vibration and state of mind.

Changes within our chemical volumes alter how we feel and, to a point, can influence how we react to situations socially and emotionally. The reaction is like tapping the sides of glasses filled with various amounts of water; each produces different tones. A cleanse similar to what I endured is suitable for anyone battling inner demons like stress, anxiety, or depression.

Having an ileostomy bag also reduced my appetite. Because I could see the digestion process from food and drink, this visual forced me to rethink what I was putting into my body. I began to drink more water and was placed on a bariatric diet by my doctors until my intestinal damage was healed. The doctors also informed me that cancer loves to use sugar, which can help build tumors. Sugar was cut out of my life after the surgery.

I find it amazing how little thought we give to our physical health and its impact on our mental balance. When we leave harmful substances behind, it makes a big difference in the quality of life and mental focus. I believe we are addicted to causing harm to ourselves because we simply refuse to give up the things we use to manage anxiety symptoms, which then promotes additional negative symptoms in the body and mind.

The brain is remarkable. It collects past data, uses it to measure the future, and helps us define our present, although we don't always calculate appropriately. Those observations and emotions represent our present. Our mind processes every emotion and sense when we sleep, but this process only completes through a specific period called rapid eye movement or REM sleep.

Each time we go to sleep, we receive mental updates and patches—like a computer—from every feeling and influence that has taken place throughout our day. This chemically stored data helps us make decisions. Our observations and emotions define our present and, in a sense, format our personalities.

We have more than enough data storage for a lifetime. Every skill, ritual, thought, habit and belief ultimately help amplify our moods, stress-free or not. At best, when we sleep, we might develop new skills, or at worst, we might create new anxieties, but like a computer, we reboot and start a new day with daily enhancements. After processing the previous day's sensory input through a REM sleep cycle, we are updated versions of our last selves, regardless of how long we slept. At a vibrational level, we are not the same people.

Whatever information we recall sets a tone openly and is the essence of our attitudes, behaviors toward others, and overall character. We use the information we have stored to resolve a variety of problems. Like a computer virus, many external influences throughout a person's day can manipulate the necessary REM sleep cycle process, and this can cause disruptions to our natural balance.

Anxiety throughout our days might compel us to retain more negative thoughts and might condition us to embrace fear more frequently. Anxiety can become our primary personality trait; processing drama always leads to more drama. We can avoid negative cycles by adopting better habits and limiting our exposure to divisive people and other sources. When we maintain our nutritional needs, avoid toxins, and remind ourselves that we have authority over what we will do next, we restore stability and balance.

Our hormones and other chemical levels in our bodies influence our behaviors too. Understanding this can be empowering and can quickly restore calm. We also need to avoid manipulative influences and all technology for at least a day or two. Some toxic items can't be avoided. My chemotherapy harmed my mental balance the most and brought many surprises.

CHAPTER 3

Faith

When chemotherapy began, I felt worse than before that large tumor was removed. As predicted, anxiety crept into my life during treatment. I knew there was no avoiding it. Once again, my mind was compromised, but there were a few tools to help me this time. Chemo made me sick, and I began to experience an unanticipated series of strange visions as well as physical pain.

I became too sick to care about those little random fears. For a moment, I thought, I can handle this. After my first treatment, I was thirsty, but to my astonishment, when I tried to drink, there was unbearable pain, like swallowing broken glass and needles. My body was also weak, and I had to learn to live part-time with an infusion machine, which controls the flow of the cancer medication. The nurses attached the device to my chest at the end of each session; I wore it in misery for a couple of days. Having cancer is a lot of work.

Most of the patients handled the indignity of each of these chemo sessions well. I was now part of this community and would learn how to take it, but I admit it was difficult adjusting. Each session usually added a new and unusual symptom. As far as I knew, there was no history of cancer in my family, so I didn't initially know what to expect from these treatments. There's no amount of questioning that can ever adequately prepare a person for chemotherapy.

I was one of the younger patients and admired the noble way the elderly patients dealt with treatment. They were fighters and were inspirational. Each session began with a nurse injecting a saline solution into the port in my chest. The smell usually made me gag. The room was always ghostly quiet, except for the occasional sound of others when they became ill; this was sadly unavoidable.

My silent boredom couldn't be broken by reading a book or listening to music, so I spent the time managing my thoughts and the strange memories that would pop into my head as I received the medication. It seemed like Mardis was sitting next to me and watching over me from time to time. With each memory, I could almost feel her presence.

The view from the oncology office was primarily sterile, but some chairs were facing a window with a view of colorful trees. I would watch the leaves fall and remember the final moments with my grandmother.

As she got closer to death, the Mardis I once knew was gone. The woman she became had a completely different demeanor. It was not because of her disease, but it felt like something more was taking place. Sure, she was still aggressive with the hospital staff, but she was no longer the same to me.

Her mental transition was astounding for a person in her condition. As a result of her change, I became a conduit for what she needed. I knew her better than I thought and far better than the hospital staff. Still, I couldn't believe I was working as her advocate when she was in the hospital. There was a spiritual transition happening in both of us during that time.

Mardis was once quite literal and aggressive in her views on matters of faith. She often read her scriptures without giving much room or consideration for metaphors or philosophical reflections, and she certainly never allowed debate. She was also unusually attracted to spiritual conspiracies.

Her conspiracies ranged from the devil leaving dinosaur fossils on Earth to confuse humanity to a preoccupation with the planet's age. I didn't realize that our past arguments would someday prepare me to translate her thoughts for others. Somehow, I knew what she needed, but my understanding was beyond family; it felt like she was in my head. Our connection was more significant than just being familiar with a person or their behavior.

• MARDIS •

Humanity might not understand all of the mysteries of the universe, but truth does exist, and that connection was indeed one of those boundless mysteries. I can't judge Mardis or anyone for their beliefs, but initially, I wanted to avoid her methods to reason matters of spirituality. I often felt like she left herself open for people to take advantage of her, and I feared I might become malleable if I stayed too long in that frame of mind. Sure, I was making excuses, but I tried my best to be as objective as possible.

I knew if I didn't attempt to see things her way, I would miss an opportunity to learn a new point of view about the nature of the universe and God. The worst I thought could happen if I didn't try was that life would continue as usual, albeit somewhat dull. I couldn't ignore our final exchanges forever. I now had time during treatment to analyze what I knew and what I could still recall from those last days in the hospital. Those memories were helpful and got me through many chemo sessions.

As Mardis was on her deathbed, I could feel her need for prayer. As I finished praying, it hit me that we never think about latency in our conversations with heaven. In every culture, prayer is instant. I considered the long-overlooked relationship between physics and faith.

I immediately began to think about quantum entanglement. Our moment of prayer was more like secretly conspiring and communicating particles in the universe. There was a harmonic synchronization at work. We were making a direct relationship and instant connection to the divine. Our vibrational change allowed our essence to reach out and contribute to the patterns, waves, and impressions with countless others, regardless of space and time, as we spoke with God.

As our similarly aligned spiritual waves connected and amplified, there were also moments of collapse. Those moments left us a space and an opportunity to be understood, but only with the determination of belief. During that connected moment, it allowed us to know our place on a waveform rippling over the universe, and we were part of something larger than ourselves.

Anyone should be capable of feeling those vibrational shifting waves. Even when we don't understand the feeling, some suddenly feel compelled to search for the source. Some might feel our presence and might be inspired to act on our behalf in some small way. Simple actions might seem unrelated but potentially cause a chain reaction that impacts everyone until those kinetic actions achieve a specific purpose. Mardis might have communicated with me in this way when she could no longer speak. I was willing and an aligned receiver who translated those needs.

It would be easier to believe that this process was divine intervention. My sister considered it a miracle because we were always at odds with Mardis. I wasn't ready to commit to anything until I understood more about what was happening between us.

The most uncertainty was whether this phenomenon and form of communication was an ability we were evolving or a gift we already possessed. If our species learn how to utilize and share discreet entanglements, there is no telling how much we could accomplish together, even to the point of moving mountains.

The Apostle Paul wrote about spiritual gifts in First Corinthians. I wondered if Mardis was using one of those gifts. I've observed various spiritual abilities in others and different world cultures for many years: knowledge, wisdom, and faith are the most obvious. Entangled connections or communication from one soul to another seemed to fit nicely into what Paul described as gifts of the spirit. I felt like I was on to something and the potential for the alignment of souls was fascinating.

I realize that most of us are rarely stewards and good shepherds in our daily lives and interactions with others. We are usually too busy or distracted, and sometimes, we are simply apathetic to the needs of the world around us. Our selfish behaviors seem to bury our incredible abilities and cut us off from our better selves. Some of our gifts might simply hibernate. There is also the potential that we don't recognize our contributions to one another and those gifts within.

Even when we are on emotional autopilot, we still manage to provide meaningful solutions and support to others, mainly when our minds are open and at rest. Minds full of distraction typically embrace posturing and empty words like "thoughts and prayers." Honestly, we know we can do better than that. Our potential can only be reached through effort, not words alone. When we act, we might also motivate others to realize their hidden strengths and potential, but action is the only way to bring about a measurable change.

Inspiring is infectious. For example, we are influenced when we see a person yawn and sense the need to yawn ourselves. We might share emotions from a distance in the same meaningful ways. When a deep emotional connection establishes, we might send comfort and strength to others.

I believe we might also feel the grief of others but mistakenly believe those feelings of sadness to be ours, not realizing that someone has shared a vibration while in distress. Sharing sorrow might be a kind of sympathetic sickness. Could this be a spiritual entanglement? Do we feel the grief of others as if the suffering is our own? I didn't know, but I maintained the idea that I was moving in the right direction for answers.

Experience proves we aren't always aware of our gifts. Accessing those gifts on demand is a struggle. While our hidden abilities are still within reach, our minds are rarely aligned enough spiritually to understand how to access that power.

Being near the right person or under the right circumstances might force our gifts to the surface. There's always the risk of being too distracted by our world to notice when our true power surfaces.

A spiritual lens is needed. It's essential to focus our thoughts, learn more about our abilities, and avoid future distractions to put any of our strengths to good use. I admit my idea of entanglement wasn't evidence, but it gave me context to understand how Mardis might have shared her thoughts with me before she died. I felt it was plausible that our communication might have been divinely motivated but also scientifically understood to a certain point.

CHAPTER 4

Connections

Many have heard stories about people who suddenly wake and feel the precise moment of a loved one's serious injury or death. Some refer to the phenomenon as the "gift of discernment of spirit." This gift is much like a psychic bond that exceeds ordinary daily instincts. Men, women, and children can feel these dreadful premonitions of loss. I would take it a step further and suggest all species have the capability.

Some speculate that these spiritual moments are initiated intentionally by a departing spirit and a need to say goodbye. Still, I wondered why some claim to have these final opportunities when others don't. Not having that last moment didn't mean it was impossible. There remains a lot of uncertainty about this unusual type of connection. I was personally unprepared to fully accept that Mardis came to me in my dream when she passed away.

With Mardis, there was plenty of warning. I knew her time was ending; it was predictable. I was aware that stress causes physical effects on the body. Tinnitus, ringing in the ears, is just one of many examples of noticeable internal impacts on our perceptions due to internal biological reactions from increased stress, but to have a final farewell before receiving tragic news is mysterious.

Those slipping into the realm of death might choose to connect because saying goodbye could be the best and only way to prepare a loved one for that horrible news. In other cases, a final goodbye might not be in everyone's best interest. Whether a person is emotionally resilient or not doesn't appear to matter. It's possible for anyone, under the right circumstances, to have these final moments. Not having those moments doesn't preclude their existence.

Beyond tragic premonitions of death, helpful and similar connections in life must also exist. Entanglements with the spirits of others could lead to meaningful changes in our lives. It could take years before we realize the gravity and impact of such alignments of the soul. I tried my best to piece together other moments in life when desperation and hope in my spirit might have cried out and aligned with others in the world.

I thought there must have been moments where distant souls outside of my small circle felt my needs and acted in helpful ways. These beings would not say goodbye; they would say, "Hello." Some might call these helpers angels. While that might be plausible, it was possible that empathetic people felt my calls for help, answered through an entangled connection, and took some small action that eventually became meaningful.

I could have taken many paths while growing up, given my unstable environment. I must have acquired help beyond the small circle of influences I had as a child. When I focused on the past, the images and memories that I believed Mardis shared with me were a lifetime of emotions and thoughts.

I continued to search for examples of that same type of connection with others, even strangers. I challenged myself to recall just one of those momentary connections and alignments. My mind drifted back to a Thanksgiving with Mardis as I considered a link.

Ordinarily, at Thanksgiving, Mardis would have fussed over food. She would only look away long enough to give us cold and angry stares of disapproval. During this particular holiday, she silently fixated on the evening news. It wasn't like her to remain quiet for so long. Usually, there would be parades of giant balloons on her television for Thanksgiving. Each of us crowded around her tiny black-and-white screen in horror to watch a news story about a place called Jonestown, Guyana, in South America.

The news reported that the Peoples Temple Agricultural Project members collectively drank a cyanide-laced grape drink called Flavor Aid, committing mass suicide. A congressional representative, his staff, and several journalists were murdered as they attempted to board their plane to leave the compound. They were intercepted and shot multiple times. The few survivors only escaped by playing dead.

The broadcast confirmed that more than nine hundred people were dead, and the footage showed bodies stacked several layers high. It was a gruesome image. Nearly three hundred children were among the lifeless in Guyana, and at the time, I wondered how many of those children were my age or younger.

I was just a kid. I didn't exactly know what a cult was, but hearing the title "reverend" to describe their leader, Jim Jones, disturbed me a lot. This tragedy was my first glimpse of how extreme psychological manipulation can impact the masses. It wasn't just the dead at Jonestown. No, everyone watching and learning about the story for the first time was affected. The report framed the event as if each person willingly drank poison, but this was murder. The difference between suicide and murder was never clear until years later.

Less than seven months after watching the horrors of Jonestown, Mardis talked my mother into shipping me off to a Bible summer camp sponsored by her church. I was terrified by the idea. Jonestown was still fresh in my mind, but Mardis wanted to see if this camp would exorcise my so-called demons and change my attitude to be more submissive and obedient. My mother loved the idea of ditching me for a couple of weeks. I thought I would end up like one of those children who died the previous Thanksgiving in Guyana.

As a child, I was genuinely curious about the world, but I didn't think asking questions could be so annoying that it would cause the adults to send me to a similar cult environment to die. I was a hostage and required to attend their camp. Running away was also incredibly impractical, and I was depressed that there was no way out and no way to win.

My mother and Mardis demanded that I stop comparing the summer camp to Jonestown. My family sarcastically said the people in Jonestown were "just a cult" and Mardis's church members were "real Christians." They didn't ease my tension in the least. Their comments only worsened the camp idea because Mardis was never a delight to be around, and if the church people were anything like her, I was doomed.

It didn't take long on the drive to camp before I began to think of this summer program as "Camp Wrath" because wrath is all that ever seemed to be on their minds. This camp felt like a carnival atmosphere, not in a good way. Immediately after arriving, an assembly took place; it was similar to a sermon. As the welcoming remarks closed, a bizarre series of demonstrations began.

Children my age began rolling on the floor, having apparent seizures in the wide-open area. Other kids were muttering incomprehensible gibberish and ululations, making noises by rapidly moving their tongues. These festivities continued for at least two hours. There were also many bouts of fainting, and it required entire teams of camp counselors to act as body catchers for the older and larger kids.

Although I was starving from the long trip, I wasn't going to ask for anything to drink or eat after seeing this behavior. My hunger would have to wait until I knew for sure the food wasn't poisoned. I also looked carefully for grape-flavored punch. The bizarre displays I was watching reinforced the idea that poison was on the menu.

I was baffled, and although I hated every moment of that assembly, I dared to ask, "What's happening?"

An older woman with a giant beehive hairstyle tried her best to explain. She spoke with a slow Southern accent but with the same energy as a train wreck, saying, "Well, this is part of our faith! They feel the power of God inside of them!" She then looked down at me with a bit of a curled lip and confusion that I didn't comprehend such a simple explanation. She was stunned that I didn't know any better, and as she looked across the assembly to another counselor, she began pointing down at my head with both hands as if to say, "We've got one."

I wondered if God was hurting these kids, and in some ways, it was a moment of isolation. I felt absolutely nothing inside that compelled me to act this way, and that was apparent to many, and the counselors looked at me as if I was against God. I was the kid in need of being saved and baptized, and I instantly became that awkward kid at summer camp.

The annual camp displayed an official theme. That year the elders' theme was: God's spiritual gifts. The camp conveniently worked in the caveat for their theme, "The Lord works in mysterious ways." This way of speaking helped them take themselves off the hook when their irresponsible actions led to disappointment and failed prophecies.

As they spoke about the gift of wisdom, their understanding was the only way. Every source of information and every opinion outside of their church was wicked. The gift of knowledge was fantastic assembly lines for cars compared to people living in huts—a central and blatantly racist talking point they demonstrated daily. They didn't hide their distrust in people of color and were openly racist.

The demonstrations of the gift of faith were the nightly sessions of rolling about on the floor. This behavior somehow proved that God personally chose them and filled their spirits with his power. Healing gifts were always hearsay stories and referenced a long string of second-and third-hand parties who might have witnessed miracles of health and cures. Even a simple tooth pain that suddenly stopped was divine intervention.

Their gifts of prophecies were usually self-fulfilling but, in most cases, utter failures. They attempted to determine the exact date of the apocalypse. Later I learned that each prediction would come and go. Their demonstration of the gift of tongues and language was useless to anyone at the camp because not a single person understood the words—they'd simply say the Lord was once again working in mysterious ways.

This crowd was attempting to summon God through each of these rituals. While I can't say that every person at this camp believed the same, I can say with certainty that the leaders of this group sought to influence each person into showing outward signs through that behavior to make the church's point and solidify their leadership and control.

Messages of love and charity were mentioned, but they were parenthetical, afterthoughts, and contradictory. Love only seemed to work if you were White, a follower of their faith, and in complete agreement with the pastor. I wanted to know more about the specific qualities of love and compassion in religion because so many people around me were just awful.

My lack of understanding meant that I was just a bad apple and incapable of learning. I felt deeply in my heart that I needed someone to explain these things to me, and I wanted to understand, but it felt impossible. I began to wonder if the only way to receive God's love was to roll around or mutter words that no one could understand, but I refused to join and thrash around their mosh pit of praise.

An elder finally told me that this behavior was, in fact, necessary. They believed that I closed myself off to God, and I needed to open up to receive God's love and power. I didn't believe them. Elders explained God would always be absent from my life because I chose to close him out of it. It wasn't that I didn't personally believe in God. I simply didn't trust those acting as spokespeople for the Almighty.

We might not always feel like we are in control, but the truth is, we are. My refusal to roll around wasn't an insult. I simply felt nothing at all. I would have surrendered my will and acted dishonestly by putting on a show. In this case, I would have submitted to those running the camp, not God.

When camp finally ended, I was so relieved. Suddenly Mardis didn't seem so bad. However, not all was lost. In the end, it took a wise old Black man to comfort me, a sad and lonely White kid. His name was Mr. Willie. He wasn't a church member for apparent reasons, but the church recognized his value as low-cost labor.

In a series of unexpected circumstances, our original driver suddenly couldn't return us home, so the church hired Mr. Willie as a last-minute replacement. Our new driver took the time to greet everyone politely. He gently reminded each of us to double-check our belongings. I felt strongly compelled to sit near this driver. After spending weeks with people who saw color as a sin, I felt a genuine kinship with this new driver.

Predictably, many of the kids on the bus were obnoxious when they noticed Willie's skin color. Yet, because Willie was so polite and patient, the kids quickly lost interest in tormenting him and huddled in the back of the bus. One of the kids in the back began showing off a *Playboy* magazine he found in his cabin as he was packing. The kids didn't want the adults to know about their prize. (These were the same kids who gave nightly dramatic demonstrations of faith by rolling around and muttering.)

Mr. Willie was familiar with their annual summer camp. He could tell I didn't have a good time but courteously asked how I felt about my visit. I confessed how confused I was about God and explained what I was thinking to him. He began sharing information about his faith, but his version was remarkably upbeat and different.

This humble bus driver shared more wisdom about the love of God in minutes than was discussed the entire two weeks at Camp Wrath. He gave off a highly infectious laugh, and he was the most interesting person I met while I was away. He provided partial answers to some of my more advanced questions, and he said that I sounded like a bit of a scientist.

As he drove the bus filled with ill-behaved kids, he managed to remain calm and found the time to inspire me on the long journey home. Willie told me that humans could do plenty when they put their minds to it, but some folks still could learn plenty. He talked about the Moon landing and shared that he watched it on television. We spoke of medicines that come from plants and many sciences. I was intrigued that this man seemed so aligned with many of the things I was trying to figure out alone.

After saying goodbye, I began thinking about his guidance, and I embraced science to seek answers to my many questions. After the trip, I never saw him again, but Mr. Willie inspired me; this entanglement was essential. This encounter was one of the first alignments of spirit I could recall that truly made a difference in my emotional state. I was now confident that there must have been others too.

Before Mr. Willie, the people I was surrounded by forced me to wonder why anyone would want to live with God for eternity, mainly because they personified God as angry and full of vengeance; it seemed more like hell to me. The church only talked about favor in paradise after people died, but the religious leaders never suggested doing good deeds or manifesting it with the life they lived. Why were they so dedicated to yelling, anger, and judgment? It was a controlling culture and full of manipulation. I wanted to avoid that culture as much as possible. Although weeks later, I gave faith substantially more consideration.

• MARDIS •

I was just beginning to get my feet wet and develop a strong interest in science. I became a frequent visitor to the public library. I was thirsty for the kind of information that Mr. Willie shared, but a storm was approaching, and the library was closed. Hurricane David was moving toward us in Jacksonville, Florida.

All of the adults were in extreme panic mode. Many people were scrambling for last-minute supplies, and Mardis agreed to take us into her brick apartment building. My sister was blissfully sleeping throughout the storm, but I was terrified. Camp Wrath might have been right, and my unwillingness to thrash around on the floor for God must have made God angry enough to send this deadly storm.

I turned to Mardis and asked what to expect when our inevitable death arrived. At the time, I was under the false impression that this storm covered the entire planet. Mardis acted as though she knew that God was only a few hours away. Her face was like a stone, and she asked me a rather weird question, "Do you have your fire insurance?"

I thought about her question for a moment with confusion. I recently heard our mother complain about paying for car insurance but was unsure if our mother paid for fire insurance. It seemed like there were more pressing worries. It was raining fiercely outside. It was the first time I'd ever seen weather so extreme and unlike anything I could have imagined at such a young age.

• CONNECTIONS •

The winds were blowing down large signs and trees. There were frequent wailing gusts and pounding rain bands that sounded like the cries of moaning spirits from hell. I was unsure if we paid for this fire insurance, why she asked, and how insurance would save us from what was about to happen.

She could see that I was still confused and slowly whispered, "You'll burn in the lake of fire forever if you don't have your fire insurance." She held up her Bible as an exhibit for my understanding and said nothing more. This exchange only added to my confusion and anxiety. Although I was naïve, I knew speaking to her more could only cause more harm than good. I remained silent and watched through a taped window as the world came to an end. As I reflect on that moment, I might have thought she was holding a book of spells and might cast one to stop the weather because back then, she did seem like a bit of a witch.

Hurricane David spared us. Soon after, we were allowed to leave Mardis and her brownstone building. Mardis gave our mother a Bible, but our mother never read anything beyond daily horoscopes and candy wrappers. I collected it, dusted it off, and began to read it cover-to-cover secretly.

At the time, the book made no sense at all. I kept reading and attempted to understand; it was a prolonged process. The words always seemed random and disconnected, but it was an advanced read for my age.

• MARDIS •

The frequently shared belief that I was mentally handicapped now felt confirmed. I simply continued to read that book without understanding. I was glad no one knew I was trying to read it because it would only provide more ammo when Mardis or my mother yelled at me again if I tried and failed to understand the text.

After four or five passes through the King James Version of the Bible, I began to memorize numerous passages. It led to even more questions about more profound meaning—those who considered themselves experts, in my small circle, never seemed to have answers. The King James Bible also increased my interest in how other cultures worshiped, and I wanted to compare ideas as I did with my friend, Mr. Willie.

Eventually, my little mind connected with some of the words, but many questions about the text were a struggle. I learned various chapters and verses, although I honestly felt more comfortable with the clarity of modern English and science books. At my local library, I began checking out the maximum number of texts daily on religions of the world and every branch of science.

As I gathered context from many more sources, eventually, I found a few straightforward religious teachings. I slowly began to understand, but I trusted that adults would have the ability to answer further questions because they were in charge of the world, not me. So many of the adults used their limited understandings to interpret what they heard, and most failed to study. The adults were so busy thumping their Bibles that they didn't have time to read them.

Fortunately, entanglements like Mr. Willie set me on a positive path. They guided me to a broader understanding of the world outside my family or circle of knowledge and limited community. With each new milestone reached and every critical decision in my life, I always found barriers, but I also found guides, like Willie, to gently help me through the next milestone.

Souls-like Willie continued to appear, became easier to identify, and occurred more frequently. Although I was considered feral as a kid, every entanglement gave me strength and balance. Many aligned souls protected me from the destruction born out of ignorance. These entanglements required specific yet whispered solutions to particular needs. Some were obvious, but many were not. We might not know where these souls lead us, but their help usually has significance, even if it takes years to realize. There appears to be harmony through each exchange.

When we understand the harmonics, our spirits can instantly reach out well beyond our physical existences and time. It might require a lifetime of trial and error, but all things in time are made clear in the end, just as Mardis tried to share when she reached her end.

A stranger like Mr. Willie might eventually discover you, offer a new direction, lift your spirits, and help you achieve much more than your current path allows. Many changes in the universe happened, which eventually led Mr. Willie to drive me on that trip home. While I admire Mr. Willie's wisdom, many others were also connected and motivated to make minor changes to their routines and lives. With every subtle change or decision made, the ultimate result was having a life-altering discussion with my friendly bus driver.

Once our souls are aligned, new doors unlock and open us to entirely new universes and potentials. Our vibrational energy and lives might change forever when we step through each door. Whether these changes are for good or bad, it's up to each of us to decide how we will contribute and live.

Every day might feel like a fight, but we'll discover better versions of ourselves with practice and effort until we find our best selves. It doesn't appear necessary to schlep ourselves around or break our necks searching for clues. As our vibrations broadcast, we seem to spontaneously connect with many living organisms no matter where they are, and this extends beyond our understanding of time.

Because we often have an impatient outlook, it can seem frustrating, especially when we don't see instant results. It's better to give up on the idea of instant gratification, especially in this century when nearly everything is within our reach or a few clicks away.

Since birth, there have been many changes to our minds and bodies. Changes to humanity have been a slow evolution from the beginning. Recent changes in how our culture connects through technology could be forcing us into a new and more rapid evolutionary cycle, making strange connections, such as entanglements, more likely than any other time in history.

We might begin to have more déjà vu moments, and other realities are bound to emerge as we get closer to a significant change in the way we think and how we act. Evolutions in all species happen when the environment forces us to adapt. Extinctions occur when a species can't keep up with rapid change. Humans have sadly been responsible for many species' extinctions, driven primarily by our greed and lack of consideration for our environmental caring capacity.

Although evolution is usually subtle, there are gains and losses in each new generation. In North Carolina, there are caverns just outside of Linville with bodies of water and trout deep inside of the dark caves. As trout moved into this environment where vision was no longer possible or necessary due to extreme darkness, they adapted; these trout became blind. Now consider how we communicate in this century and its impact on us.

There is no ignoring that technology has caused numerous changes in our social behaviors. We use technology to gather information without questioning accuracy. At any given moment, most people in any situation are staring into tiny screens and connected to electronic devices, and we don't look up often enough.

Our impatient attitudes and desires for information without careful reflection can lead us to trust terrible guidance, similar to the type shared at Camp Wrath. In our current state of development, we are slowly becoming a more reactive species than a reflective one. We quickly turn over our better judgment and fear stimulus in such a careless manner to unknown sources online.

As we grow more complacent, we begin to open our minds to further risk of manipulation by those who choose to create chaos. As our behavior continues to accept this imbalance, mental atrophy reduces our ability to behave with honor and civility, and eventually, we might become more like pack animals than people. However, these risks can be mitigated by turning off digital influences more frequently and doing more internal reflection.

Consider how birds have adapted to the modern world, thanks to the noise generated by progress, planes, and automobiles. Mockingbirds, for instance, in the suburbs and cities now mimic the sounds of automobile security alarms and are forced to have earlier singing sessions while searching for their mates. Birds compete with noise pollution from traffic and civilization in general, as they struggle to adapt.

    With each new product we purchase, we destroy a little more of our environment. We ignore the destruction we are causing. Even lifesaving medicines are destroying ecosystems. The chemotherapy saving thousands of people like me each day can ruin habitats quickly. Whether we want to realize it or not, we are all connected in this world. The Earth might eventually survive, but humans most certainly will not unless we restore balance and acknowledge that we are all connected. Not considering the future makes humanity a lot like those trout in the cave; we are blind to many things.

CHAPTER 5

Hypocrisy

Cancer brought me close to teams of nurses, pharmacists, and doctors. It was refreshing to be in an environment filled with trained professionals. We were battling to understand my condition and resolve it together. I embraced the hope that soon, I would be cancer-free. An atmosphere of science always gives me peace of mind, but I also have a healthy respect and love for theology.

Like many people today, I began to see faith as a form of manipulation. I learned that religion was not a problem but the hypocrisy of the messengers, on the other hand, was severe. Unlike scientists, people in the business of faith often ignore the data. Beautiful guidance and grace are frequently twisted and spun to match personal narratives and biases. They turn blind to spiritual guidance to achieve selfish and personal gain; this is not faith; it never will be. While I'm certainly no saint, I do embrace the many sacred lessons from many theologies that encourage the good we have within our spirits.

Growing up, Mardis insisted that we attend her church, and our mother agreed because she always needed her alone time. These church services focused on the apocalypse and were far removed from charity, lessons of mercy, and commandments to love one another.

It's easy to characterize her former congregation as a Jones-style doomsday cult, but Mardis swore that wasn't the case and would threaten anyone who suggested otherwise. The church that Mardis attended was held together by the collective despair of attendees and their tithings. Her pastor was always happy to accept every penny added to his collection baskets. He was a highly confident man but would frequently lean in so close you could almost hear the sound of his five o'clock shadow growing. His greasy hair usually dripped on the clothing of those he spoke with in this way for too long.

This man displayed unusually crooked teeth, covered in a layer of filth with the same consistency of Parmesan cheese. Although his breath was putrid, he loved to speak and never seemed to stop. He would loudly shout that his teachings alone were the only way to handle this horrific world and be close to God through his sermons. He also frequently made irresponsible predictions about the date the world would end.

• HYPOCRISY •

Mardis forced us to spend more than a year attending his services. We participated during the week, at night, and every Sunday. She always ensured that we carried coins when the baskets for tithes came around. It was blasphemy if we didn't give to the church, at least in her eyes. She was a highly dedicated member. However, it always seemed like the church was doing far better than the entire congregation.

Each time we arrived, Mardis and the other ladies at church would gather and share various stories. Their usual gossip was about the pastor's divorce and his new luxury car purchase. The pastor was incredibly proud of a brand new Lincoln Continental, and he brought it up in nearly every service for about a month after he purchased it. He claimed that his luxury vehicle was a sign of prosperity for everyone and guaranteed that his flock would see this same type of windfall in their lives.

There was always a provision for his promised financial prosperity: "Continue to give generously to the church." Even at a young age, I was confused about why anyone would need a new car if the world were ending. The pastor once quipped about driving with Jesus in style when Jesus arrived.

Mixed in with his contradictory sermons, opinions of the nightly news, and car updates, the theme about the end of time remained his top priority. I heard seemingly endless tales about how the world would end, and his constant prattle might have contributed to my nightmares. I often dreamed of people bursting into flames and disappearing. The remaining people of the world would hunt me like an animal to torment me for my insubordination.

His end-of-time dates kept getting pushed back with each failed prediction. He often claimed God would speak with him and task him with a new job to save us all for another day. His sermons continued in this manner as long as we attended, and they never got any better.

Gathering with the other children in Sunday school at this church was painful. The lessons were arts and crafts. We cut paper and added angel wings with paste to toilet paper rolls. There were no religious discussions in these small settings, and the adults held other talks that we couldn't hear. At the time, I thought the kid-free talks were when they did the serious discussions.

After Sunday school, I made a critical mistake by openly admitting that I didn't enjoy the church. Disclosing this in front of Mardis's gossiping friends was a terrible thing, and it brought Mardis a lot of personal shame and anger. The pastor didn't like my statement either, but he comforted Mardis and casually asked Mardis if I was mentally retarded. I didn't blend in with the other children in Sunday school or their previous summer camp retreat.

The pastor's statement was another reason to dislike me on Mardis's growing list. Later, my mother gave me a severe beating for being stupid and worthless. I tried to explain that I wasn't against faith at all, and I wasn't evil, but I just couldn't take it anymore. I couldn't listen to the constant shouting about the entire world coming to an end. No one understood why this bothered me so much.

The church members didn't seem to understand that children look forward to their turn to become adults, but according to that church, there was no promise of ever living long enough to become one. Fortunately, that year we were moving farther away. I was no longer required to attend the services. Even if I were miraculously willing, it was simply no longer possible, given how far away we moved that year.

Over many years, I've joined in prayers with others at synagogues, mosques, and temples. I've visited these places for various research projects and honored many customs that were not my own. Although my study in such sites offered a clearer picture of each culture, Mardis and her congregation would have said that I might as well practice black magic and voodoo dancing. They would have seen my behavior as obscene and with absolute disgust.

Curiosity and a desire to understand the world have taken me to fascinating places. I've been to smoke-filled dens to have vision quests with shamans. I've explored the totem world to meet my spirit animals, and I've entered musty libraries, where I visited and interviewed mild-mannered academics at universities. I have tried for decades to learn as much as I can about the meaning of life. I hoped to learn more about what waits for us all when our time finally comes to depart the world, so I continued to search for answers and ask questions.

I've studied ancient civilizations and theologies and spent countless hours doing field research, interviewing believers and nonbelievers in various faiths. Curiosity is a blessing and a curse because it rewards us with high mileage and loads of maintenance. Still, in the end, the fresh perspective and broader understanding of the limits of humanity are worth every minute and ounce of trouble.

I've discovered that looking to religion for direction can be an excellent place to begin and, at times, somewhat rewarding. However, not every person claiming to be of faith is qualified or capable of providing meaningful spiritual guidance or wisdom. Many in a religious leadership setting are simply in the cash role and have no problem saying untrue things without considering consequences. Those in the business of religion mostly do it because manipulation is their craft, and they are skillful.

It might be surprising to learn that even those people selling faith have developed a sense of entanglement. Sure, they lack the genuine feeling or desire to help for the sake of usefulness but say they can feel the needs of others and use those feelings to influence. They target those in despair

• HYPOCRISY •

and siphon money from the desperate who only want relief from their pain. Everyone should understand faith is free. Many religious leaders will attach a price tag for their claimed relationship with God, but even the devil can quote scripture.

Independently, if we choose to read holy texts and use them as a guide, this can be an excellent way to find peace for many, but the sacred text can also be challenging to read and understand. If we turn to others without genuine knowledge and wisdom, we quickly learn their motives usually involve money or ego. In many cases, we end up broke and sometimes worse off than before searching for guidance.

We should never allow anyone to convince us that salvation through love is sold separately or made in installment payments. Mardis cared deeply about her faith and was willing to give everything to demonstrate her devotion to her church and her gossiping friends, but this wasn't necessary.

Unexpected encounters just happen, and those willing and capable of providing meaningful assistance will never request a dime from us or a favor in return. Many will still do good deeds for no other reason than helping others, no strings attached. Proper alignments begin within us, but religious leaders busily begging for cash will force a connection and always desire to be the hero, expect constant praise and tax exemptions.

Sharing and giving to others is essential, but sometimes those seeking help are only a few steps away. Simply throwing tithe money in a basket does little when we overlook or ignore those closest to us in immediate need. There is no requirement to be a member at a specific church, and there is no scorecard, but we must have a willing heart and a focused mind.

Our ancestors' guidance often reveals how decent we can be to one another when we make a serious effort. Far too often, those beautiful teachings become corrupted by those searching for a stooge or a new private jet. These ancient lessons are essential parts of history, and they should be cared for with honor, not treated like poorly created infomercials to raise money "for God."

This gross misuse of spiritual trust and the many examples of manipulation through religion have negatively and significantly impacted our entire world; it's staggering. Anyone can act godly, but faith without an honest effort to help others is nothing more than empty words, and an empty faith is dead on arrival.

During a moment of entanglement, we must feel a calling inside to assist someone in need. As those feelings begin to energize us, we might feel compelled to take action. Taking action of any kind usually helps others more in minutes than most churches will help in decades through their assembly. When our hearts are in the right place, there must be a compelling feeling that will grow, and the vibration becomes louder in our souls, encouraging us to act; the rest is entirely up to us.

I have learned to recognize offenders of faith by their messages of division, their unwillingness to show decency to anyone different, the silence when the world needs moral clarity, or how often they ask for money. Those who remain silent are usually the worst because they are often more afraid of becoming unpopular. They are far more fearful of tarnishing their funding sources through action and care less about morality, more about power and finance. We typically learn that they have the darkest secrets and are unworthy of trust.

These imposters range from bad to worse. They rarely act with the integrity required in their roles and make bold cases for the financial costs and burdens of spreading God's message. If a sermon on the mount was good enough for Jesus, it should be good enough for them. Transactional compassion has been around for a long time; even the apostles in the Bible knew this.

In the days after the crucifixion of Jesus, the apostles struggled with grief and to keep their religious movement going without the constant physical presence of their teacher. Others began to surface and take advantage of the religious movement that would someday be called Christianity. Many performed deceptive magic and legerdemain with their accomplices hidden in the crowds during that period. These men were mainly opportunistic pickpockets. They performed their magic for money.

One of those menaces was a man called Simon Magus, a magician. He was in awe of what the apostles were doing and wanted to learn how they accomplished such marvels. It is unlikely that Simon intended to live a life of love or, for goodness sake, if he acquired such ability. Simon attempted to hire the Apostle Peter to teach him to perform miracles.

Peter knew that Simon Magus was only interested in scamming the weak in the name of God. This encounter is in the Bible book of Acts: "But Peter said unto him, 'Thy money perish with thee, because thou hast thought that the gift of God may be purchased with money'" (Acts 8:20, KJV).

Later Apocryphal writings explain how Simon Magus eventually got what was coming to him. Simon Magus levitated in the air for a crowd as the Apostle Peter looked on. Peter was able to see demons suspending Simon Magus in the air by his arms. Peter simply commanded the demons to leave in the name of God. The demons feared Peter and fled. Simon was dropped to the ground, shattering bones, and died later of his injuries after the fall.

Like Simon Magus, many modern evangelicals would have us believe that we must contribute financially to speak with God or to receive blessings and miracles, but money is useless to God. Many antipopes in the Middle Ages would use their positions in the church for spiritual extortion, claiming to free damned souls in purgatory for a price. This practice of paying for religious favor is now called simony. Each time an on-air evangelical asks for money and promises prosperity, you are witnessing simony.

My constant exposure to the negative side of religion led me to study, and that effort brought me to an understanding. I developed a deeper love and respect for faith and an absolute grudge against those who misused it for personal gain. I felt an obligation to protect sacred lessons and not exploit them. I realized that there is much more to faith than those pandering pastors could ever ad-lib. Mardis fully understood this side of me before she was ready to depart this world.

Later I thought if I managed to survive cancer, it would be nice to help others avoid the manipulation that Mardis might have been exposed to throughout her life. It costs nothing to be a decent person, but paying for favor and prosperity will always come with a heavier spiritual and financial toll than most people realize or can satisfy.

Performing miracles should always be for the sake of helping out of love, never for repayment or to be used as leverage. Internally we'll feel that urge at the right time to help, especially when our spirits are aligned. Sometimes all we have to do is open our eyes to see those in need. When that vibrational calling encourages us to take action and help others, we should never use their moment of weakness to gain an advantage or seek control. If we behave that way, we are no better than Simon Magus.

CHAPTER 6

Answers

At the end of her life, no matter how much she tithed, Mardis was alone. Dementia spared her from the revelation that everyone she once praised or showed kindness and devotion to wouldn't be with her, not even in the final minutes of her life, but love prevailed. Even at her eleventh hour, her isolation shattered, thanks to my sister's efforts; we both agreed that no one should die alone.

Still, I would have never imagined that I, of all people, could sit by her bedside to hold her or to read to her until she could sleep. Although she often woke and forgot who I was, I didn't mind. She saw me as someone different each time but always peaceful and worthy of kindness. I struggled against those brief moments, and it was hard accepting such honor; it was also confusing. I was grateful to be there for her and became emotionally anchored to Mardis.

Usually, I wouldn't think of my emotional responses with such care; I didn't believe I could cry. My emotions seemed to be hot or cold or, in simplistic terms, good or bad. I was usually too busy to notice all the nuances and hidden details or depths of every feeling. There were just too many variables to keep track of and not enough hours in the day. The moving way of Mardis shared her life with me left me in wonder.

As I attempted to find a method that could easily explain how her spirit shared her thoughts, especially in her mental state, I tried to recall every philosophy and scientific approach to understand. I might have simply been in shock, but I wanted to learn how she managed to do this or if it was simply all in my head. I made excuses ranging from hunger and fatigue to simply misunderstanding. Yet, the thoughts she was sharing continued to the point of discomfort.

Whatever was happening allowed me to feel and understand what Mardis needed but couldn't communicate to the hospital staff. I felt like I was downloading many of her memories. Although I was determined to help her feel as comfortable as possible, I became compelled to act without thinking or the need for spoken words.

Mardis and I were different personalities, but our spirits were of the same substance and somehow synchronized like particles. We were two sides of the same coin spinning on edge. As long as this spin lasted, she could share her thoughts and receive mine. My only desire was to show compassion, but our shared connection somehow continued long after her death. During my cancer treatments, the memories and coordination returned fiercely and convinced me that our interactions were outside of mere considerate behavior.

Thoughts continued to come and go in waves, often amplified and intense. My heart and mind were conflicting, and I felt like I was me and not me at the same time. Each sensation felt like they were somehow part of the existence of everyone that has ever lived or will live. Our spiritual coordination and collaboration behaved one way when I purposely tried to focus on those thoughts and another when I was trying to mind my own business. The only remedy was to shut up and let it happen.

Was it possible that consciousness and feelings were instantly shared between humans in this way? Entangled vibrations of others might be on emotional frequencies where we potentially exchange optimistic needs and hopes or even our negative fears—anything.

Aligning with other souls, our shared thoughts and needs from the combined emotional vibrations might inspire or guide us. We might feel compelled to act, but the choice to move is ours to make. Engaging in an entanglement could be as simple as starting a conversation with a stranger, smiling at a person passing by, or connecting with others to share in their hopes and dreams. Even the slightest action can have enough resonance to shift the universe. Ultimately two souls will begin to transfer information, but the similarities of thought would be immediately felt and surprising enough to some people to directly impact their behavior.

This concept would likely be similar to how a service dog can detect a pending seizure in a human. A change in the handler's vibration is detected, and the service dog responds. I understood requests and responded with assistance, much like a service dog would but more. I could see flashes of images like an old film of people, places, and things, but it was all moving too quickly for me to make any sense of it. I only understood what Mardis needed and what I needed to do to help her. There was no other context, but out of compassion, I took action.

As we receive vibrational patterns, we might feel inspired to focus on the sources, and we start to feel that something is different in our spirits. We can't know what is truly possible with these emotional vibration connections until we agree and connect. When we find the sources, we find one another, and the possibilities for life can seem limitless.

These connections might lead to long friendships and relationships, from chance encounters to hearing the right words at the right moment. As time passes, each relationship becomes more meaningful, new doors open due to those aligned histories, and more entanglement connections emerge. The vibrations of our atomic structures could change and allow us to slip into new versions of ourselves and new universes of existence.

Mardis and I might have also made emotional connections that simultaneously connected us with countless souls. Vibrational alignments likely happen more than we realize. A feeling that we know a person because they seem incredibly familiar appears to be typical of this type of entanglement. Other links, such as déjà vu or two people saying the exact words simultaneously, also appear common. Each of these strange connections is like one of many threads that hold our universe together.

Entanglement is likely beyond synchronicity or a series of random coincidences. An entanglement would be the functional changes to the vibration of spirit broadcast on a transmitting engine, powered by individual spiritual needs and desires of a specific nature. We are essentially asking a question while standing in the enormous hall of our universe. As the souls of others in this universe feel and hear our questions, they respond, creating new entanglement waves and distorting what we think of as the norm.

The interactions and changes around us might seem random from a certain point of view, but we see these effects as causality, not a coincidence. Synchronicity might need a different point of view to understand what is happening. Those without context might only notice the aftershocks of an entanglement exchange by those on a different frequency. From the point of view of synchronicity, changes in our world might appear random. Observation from that level is only understood with the whole spiritual context when those with that line of sight adjust and join that harmonic discussion.

We understand and relate more frequently when we are empathetic and free from hostility, and it appears that we are strongest when we simply mind our business and our minds are at their clearest. When our minds are still, an attraction amplifies, and new connections gravitate toward us for specific purposes based on our mental frequencies. We become part of the beautiful vibrational songs shared with countless souls in the universe.

Sure, I admit this might sound "hippie-dippie" to many people, and they'd be right to feel this way. Not everyone will understand this concept or even believe it, but that is ok. Consider this, and it's essential always to look deeper. A fresh outlook and consideration of the actual needs of others often encourage a positive approach to life in general.

Positivity only gains strength if we nurture one another with considerate behavior and other meaningful ways. On the opposite side, using this method to discourage others will only carry souls plunging into the darkness of negativity and unhappy existence. These concepts are similar to heaven and hell.

• ANSWERS •

Try to visualize a droplet of water and the ripples that begin to move out as that single drip becomes one with the larger body. You're that single droplet; your emotional reactions are those ripples. As you become one with many in the larger body, your initial action causes an enormous reaction on the surface, which we think of as the universe.

As those waves move and those reactions are shared, others will receive something new in their spirit that might not have been there before. New patterns and waves emerge as those ripples reach others who feel your needs and respond. As ripples collide, they amplify, and new connections occur. Some pattern designs will be perfectly aligned, and the world around you might subtly transform as the many new waves change shape, creating other unique patterns.

This experience might lead us to new solutions to problems, there might be more opportunities, and our thoughts will fill with clarity, but for a specified purpose. When we align, we might share our wisdom and energy, even if we are initially unaware of that transaction. The result produces an element of binding harmony and charm related to the aligned assistance and needs.

Although many of us might still focus much more on the negative in our world, there are decent and genuine people who are willing to share "good vibrations." Because negativity is divisive, it's difficult to hear those we should listen for over the noise and chaos in the world around us, as it attempts to cancel out our connections with one another.

Cleansing our minds and bodies is a good way of clearing up the background static and distortion or noise caused by the disorder of our world. Clarity and focus begin within us when we work together. As our spirits coalesce, the vibration grows more intense and reaches a chain reaction that can be extraordinarily powerful.

As Mardis would think of a request, I seemed to instantly anticipate it, understand it, and take action: food, crushed ice, help, and bathroom. I might never fully know why we made this connection of thought, but I realized she needed me. Closer to her end, her dementia gave way to aphasia and made verbal communication with others impossible. Still, when she lost the ability to speak, our connection only seemed to grow beyond mere crushed ice.

Mardis shared her entire life with me within just a few moments. I gained a deeper understanding from her lifelong struggles and tried to think of the event as an alignment of two life forces and histories. Mardis gave me a spiritual context, where I could understand her, and she could understand me completely, even though our mutual understanding was for fractions of a second before her life ended.

The understanding that came from our entanglement was impactful. I began to identify parts of myself once thought lost.

• ANSWERS •

I discovered a new determination for a new direction in life. She also helped me realize that I needed to see myself differently, beyond the projections and opinions of others—even those negative ideas that she once planted within me long ago.

A part of me still realizes our entanglement wasn't magic. It was easier to understand, as we were both in the same room. I understood her medical condition, I could tell when Mardis was uncomfortable and needed specific help, and I attempted to anticipate those needs to make her more comfortable. Although acting this way felt like more than courteous behavior, events like these would happen more often than with just Mardis. Being with her was a catalyst that showed me there is far more to life than seeing, hearing, and believing. We haven't even begun to scratch the surface of our potential.

As I looked back, I discovered connections were more common throughout my life. I tried to remain objective and cautious and attempted to distance myself from fringe ideas, but I recognized many gifts in so many people. Because we are all so busy trying to keep our heads above water, most of us never realize our potential and are almost robbed of the opportunity to be the best we can be.

I remembered one visit where I was waiting just long enough to ensure Mardis was peacefully sleeping before I called it a night. As I sat listening to the beeps from heart monitors and the muffled alarms down the hall, I began thinking about the patients in nearby rooms. I wondered if they were ordinary families who shared love and genuine concern for one another, free from guilt and scandal, or were they more like us?

My sleepy mind drifted further, and I felt Mardis was guiding my spirit to many nearby hospital rooms, where we would feel regret in some rooms and boredom in others. It was unsettling, and it left me feeling like I was dead and might never make it back. I dealt with this memory by reminding myself how tired I was that day.

I considered the event as fatigue. Before drifting off to sleep on that visit, friendly nurses walked in for their shift change. They whispered greetings and asked if I noticed any difference in her condition. After what I experienced, I should have yelled, "Definitely!" but I was certainly not going to admit anything about that event out loud. That night, I drove home thinking about how bizarre the visits were becoming and felt there must be a reasonable explanation.

Our connection continued to unravel, and my thoughts always seemed full. Mardis shared some terrifying ideas and other ideas laced with ancient wisdom. Each day thoughts appeared to wear a little more in my mind. I tried to convince myself that these ideas and thoughts were from self-reflection, but they continued and didn't shut off.

• ANSWERS •

Even after leaving her at the end of each day, I felt trapped. Although I don't believe Mardis was fearful of her fate at this point, I think she might have feared for my spirit and my soul's well-being. She was doing her best to guide me with every ounce of power she could spare.

Mardis was always bitter in life, and I understood that, but becoming bitter myself was a choice. It was also why I always felt so dreadful, and I made that choice alone. Understanding this and acting against that inner compulsion to act negatively only to blame a villain was eye-opening. Our reunion in the hospital setting helped guide me to that basic understanding.

Our connection left me feeling shameful in many ways, mainly because it took me so long to realize how I allowed others to influence me when I always could make a choice. I also understood that regardless of the negative people in my life, they were not the gatekeepers of my hell; it was always me. Only I could walk through the exit and leave my madness behind.

Her weird and wonderful cosmic messages might help many more people avoid poor behavior, but not everyone will be open to change. We have an abundance of control over our lives and how we live them, so it is up to each of us to make every single opportunity count. Individually we can set better examples for those who feel they are lost. However, our universal duty will always be to inspire.

CHAPTER 7

Isolation

I suffered from a variety of reactions to chemotherapy. I usually bounced between throwing up and passing out. I also experienced strange lucid visions. It was necessary after each session that I make it home as quickly as possible, but on one visit, I forgot how to get there and became lost. Some patients have called this kind of forgetfulness during treatment "chemo fog or chemo brain," and no matter how hard I tried, I couldn't remember how to get home.

I decided to use my phone's Global Positioning System to navigate. Rather than taking a direct route, I must have chosen the longer scenic route in my confusion. The absolute trust I was giving to a program for essential instructions was mind-blowing, and I felt worse by the minute.

If I am willing to trust my phone for guidance, how often have I been submissive to people over other important matters? I have often felt lost, but how often was this due to trusting the wrong people, especially if there was a history of poor behavior. I couldn't count the number of times I was reliant on bad advice.

When I finally arrived home, I considered my life. The only thing I knew about myself was a veneer of opinions that belonged to other people, but a part of me relied on those opinions. Suddenly, I started to blackout as I struggled to stand, but there was a spark of absolute understanding, although it was brief, I realized what Mardis shared with me. The moment the epiphany happened, it evaporated, and my heart started to race.

I began to notice a golf-ball-sized lump under my jaw beginning to grow slowly, like a water balloon. I gently touched my jaw to see if the spot was a hard or soft mass; I instantly regretted that decision. My heart rate became so high that I thought my heart would burst, and I began to drip heavily with sweat; I was saturated.

I panicked. I struggled to make it to my couch and call for help because I felt like I was going into cardiac arrest. Fortunately, the symptoms began to fade, and the lump on my jaw also began to shrink. The infusion pump medication pooled in a specific region on my jawline and filled with toxins. Even the slightest pressure when touching that lump was enough to fill my veins with the drugs rapidly. I passed out and slept for hours.

• ISOLATION •

When I woke, my previous foggy revelation about Mardis remained elusive. I could feel the answer was there, and it was just as real as the bulky infusion pump I was carrying. At the time, everything seemed so clear, but I simply couldn't retain the thought. I clung to even the slightest memory, but sadly only vague fragments remained.

Over and over again, I went through everything I thought before I passed out, but my mind kept returning to a simple thought: pettiness. I tried to move past the idea, yet it persisted. I was winding up with frustration, out of touch with whom I was because I always relied too heavily on other people.

I needed to learn more about my spirit and meet my authentic self for the first time, free from other people's lenses. In my case, the bar was low; I was convinced that I was trash. When I looked closer at my life, I recognized it was different than what I was conditioned to believe. I achieved a lot without realizing it. I earned degrees in cultural anthropology and history. I was also an engineer and steadily worked in many scientific fields. I was even fortunate enough to be invited to Abu Dhabi and Dubai to discuss emerging technologies in the United Arab Emirates. Still, none of these milestones in my life mattered to those who wanted to see me fail.

All it took for negative people to disparage me were social media photos from Abu Dhabi and Dubai, where I was smiling and happy. In one photo, I was standing near the Grand Mosque with my host, and this setting was just too tempting for gossip to Mardis and my mother. Immediately Mardis and my mother created a narrative where I was somehow a priest for Al Qaeda, an agent of the devil. This manner of painting me was just one of the countless battles with them and how I was always unfairly dishonored.

Suddenly, I remembered what Mardis said to my sister before she was taken to the hospital years ago: "There is just no pleasing her." Although I wasn't with Mardis when she spoke those words to my sister, I could hear her say it, and it gave me chills. The hair on my arms and neck stood up. After receiving this message, I stopped caring about the thoughts and projections of others, and it was liberating. Later, I became my own worst critic, and while my views of self might have often been harsh, I was at least realistic.

I was still in bed. I tried to open my eyes but was still fuzzy-headed from the infusion pump drugs. When I could finally see, I was back in the hospital and sitting in that cold chair near Mardis. Part of me wanted to believe that I was taken to the hospital, but this setting wasn't natural. Once again, I was reliving a moment before Mardis died. I was fully aware that I would forget much of what happened when these sensations ended. I tried to find a way to recall as much detail as possible because, like a dream, only fragments typically remain after we wake.

• ISOLATION •

Slowly her room was coming into focus, and I began to notice Mardis, but the room didn't look quite right; the lighting was a little darker yet somewhat blue. The room shone with a soft quality. It was the kind of ambient lighting usually caused by a rainy day. I fought the hallucination, but there was a moment when the thoughts came pouring in again. I couldn't distinguish between her thoughts and my own.

As her face came into focus, Mardis slowly turned her head and looked in my general direction from her hospital bed. There were oxygen tubes around her face, and I could hear the high-pitched tones from her monitors. She was still holding my hand. She weakly nodded her head with approval, gently squeezed my hand, and seemed to look the other way to sleep.

I held her hand close and as tightly as I could without harming her. A wave of approval came over me, and I accepted that I should never trust negative minds to determine my path forward or guide me. The flood of thoughts felt like a series of rules or commandments. Mardis revealed that inner demons are desperate to cut us off from one another, blinding us to all things good. They manipulate us to keep us ignorant and so selfish that we never reach out to discover one another.

Love is virtually everywhere but hidden in plain sight by the daily negativity surrounding us. I understood that we should carefully listen for souls reaching out, even when we don't feel deserving, and I began to hear the sounds of music, but there were no words.

I could hear long and short tones played like perfectly synchronized stringed instruments in a symphony mixed with sounds that I have listened to from data gathered by radio telescopes. The music began to fade, but the commandments continued as images of people, places, and things moved so fast. I closed my eyes because I was becoming uncomfortable, and suddenly everything stopped.

I opened my eyes again, but I was at home in bed. My phone's Pandora music app was open and playing a style of music similar to what I heard moments ago. It must have just been another weird reaction to chemo or another odd dream. Still, I carried something substantial with me this time, an understanding and a determination to meet myself for the first time and avoid allowing others to drag me down with their harmful opinions and ideas.

I'll never please everyone, and I shouldn't waste trying. I knew that I should conserve my energy and use it to do good for those who need help rather than wasting that positive energy on those who never really cared or deserved it.

The following Friday, when I made it back to the oncologist to return the bag, I explained what happened. The oncologist gave me a stern warning because the reaction I was reporting was similar to cardiac arrest. I was lucky that time

• ISOLATION •

because I could have died, but I wondered if being so close to death was the reason I understood what Mardis tried to share and if I only felt the clarity while at the edge of life.

CHAPTER 8
Charity

I was a little past the halfway mark for chemo. I knew I was getting closer to the end of these awful sessions, which excited me, and I hoped that these bizarre visions would also stop. I arrived early and found one of my favorite chairs with a view of trees outside the office. Chairs quickly filled up, so I was lucky that I got such a lovely spot.

A mother and daughter walked in together, searching for a place to sit, and chose a spot near me. The mother was elderly. The daughter was closer to my age and stayed with her mother for the entire session. Each session took a little more than an hour, but she never left her mom's side.

I listened to the two ladies carry on with nervous banter, knowing that any one of us could get violently sick at any moment. The mother suddenly remembered that she needed to do some laundry. The daughter demanded her mom stay away from doing anything physical and insisted that she rested. She promised her mother that she would take care of her laundry. The elderly woman just smiled and said, "You're such a good girl." Their situation was so different from Mardis and my mother.

The last chair was about to fill up in the room, and it was a man. He was by himself and looked miserable. He was thin and worse off than most patients and reminded me of someone I once met, but I couldn't recall whom. The new patient took a seat by one of the nurses who usually prepped my chemo port.

Each time I got this particular nurse, I received a specific drug that only she would deliver. This drug made me tired and usually sicker after the session in the chair, and I was always worse when I got home. I knew when I saw her that the session would be a hard one to endure. She flushed my port with saline, and the usual awful taste filled my mouth. She started my drip but gave a separate injection. I quickly became drained and passed out.

Somehow, I could still hear the mother and daughter carrying on in the chair nearby, even though I was groggy. I kept thinking of the mom trying to do things for herself and the daughter taking them on instead. Mardis should have had the same courtesy. It is hard to understand why Mardis would have allowed herself to become a servant to my mother, and I wondered how much longer she might have lived if she didn't have to endure so much unnecessary work.

• CHARITY •

There was rarely a shortage of communication with us because we were once daily targets of their loud attacks for the audacity of living. We chose not to participate in their abuse. Our mother might have kept Mardis's illness a secret, or Mardis might have chosen to do so because of her pride.

If we had known earlier that Mardis needed help, we would have. I can't say that support would have been instant. I remained guarded and extremely cautious, but my sister felt a deep, compelling need to visit. Meanwhile, in a nearby chair, a different type of relationship existed between a mother and daughter, and it was beautiful.

Mardis often possessed too much pride to ask for assistance when she was healthier. She was also highly opinionated about anyone reaching out for help unless it was a person she knew well. Mardis would never give to a person on the street. Mardis warned us to be fearful of begging. Charity only went to those she deemed worthy or those she approved of, and any others were called "hippies, beatniks, and lowlifes."

There were two long-held maxims that Mardis and our mother loudly and frequently shared with their victims: "You made your bed, now lie in it," and, our mother's favorite, "You play, you pay." They thought of people in need as lazy, addicts, or sinners. They considered people in need unworthy of assistance, regardless of their circumstances, and never cared about the complexity of others.

• MARDIS •

Neither Mardis nor our mother showed compassion unless there was something in it for them. When they did help, they required constant and unending praise and the expectation of someone owing them something. That is not to say that Mardis or our mother never did an ounce of good. But when they would help others, there was always a guarantee that strings would be attached.

Sometimes pride keeps us from finding the help we desperately need, especially when we fear being taken advantage of at a vulnerable moment. Mardis finally allowed others to demonstrate that payment for assistance was never necessary when she allowed us to visit her in the hospital.

While I generally don't like to discuss the help I have given to others, there are rare times when sharing a moment can be meaningful and necessary. Sharing certain moments can provide adequate warnings and help avoid future and highly unnecessary suffering. Assisting others should never be used to satisfy a crass inner desire to brag, so we must be cautious about what we share and the reasons for sharing.

In my early twenties, I was broke. Budgeting my income was not always easy, so I stopped carrying cash. One day, I stopped to get fuel, and I noticed a man digging through a nearby coin-operated car vacuum. He managed to find a way to open the tank and searched through it for coins. As shameful as it is to admit, I was privately grateful that this man, not me, was digging through garbage for money.

• CHARITY •

He began to walk in my direction, and I knew what was coming: the request for help. For a moment, I was relieved that I didn't carry money and could truthfully say, "I don't have any; I'm sorry." It wouldn't be a lie, and I hoped that I wouldn't feel guilty for saying no to a person in such a crisis, but his dilemma seemed far worse than mine.

He asked for the handout, and I politely gave him the bad news but still felt horrible about it. He said, "Thanks anyway," politely waved, and walked away with an unusual gait. I recognized his distinctive walk. He was suffering from malnutrition and scurvy. He was gaunt, his clothes were hanging off him, and there was a reasonable chance that he was severely dehydrated.

I felt a powerful internal calling to go back for this man as I drove away. It wasn't like a voice. It wasn't words at all but an incredible instinct and pull to find him again; it wasn't something I could easily ignore. Call it guilt, but there was such an intense urgency to circle back and see him as quickly as possible that I dropped what I was doing and went back.

As I circled the traffic light, I spotted him again, sifting through another vacuum on the opposite side of the street. I walked into the store, used the automatic teller machine, and bought him a few things. He recognized me as I exited the store, and he knew my answer if he asked for help, but this time, there was cash in my pocket.

MARDIS

Although I purchased a few items for him, something inside me still demanded that I do more. Rather than give him cash and the newly purchased treats, I said, "Let's get you some food. Hop in." Mardis would never have approved what I was doing, but it was abundantly clear that this man needed a decent meal and calories fast.

He was in such awful shape my first instinct was to find him a doctor. He pointed at a nearby McDonald's, but we couldn't go inside because sitting next to him took my breath away; he smelled like rotting meat. I asked where I could drive him to enjoy his meal in peace. He just pointed in a general direction, and I began to move.

As we drove, he began to tell me a little about himself; he was a genuine hobo. He introduced himself as "Royce." It was hard to understand his words due to his lack of teeth—only two or three were visible. As he devoured his food, he told me this was his first hot meal in several months. I warned him to slow down. I didn't want him to get sick. It must have occurred to him that no one would take his food away, so he finally slowed down.

When my vents were directed at him, my air conditioner seemed to reduce his heavy odor. Before I felt that much stronger urge to buy him a hot meal, I purchased orange juice, water, honey buns, and aspirin. I added change to the bag and tied the bag together so I could leave him with a care package. I thought it might help him avoid dumpsters later.

• CHARITY •

I hoped there would be someone in his life to call for help, offering more assistance than I could provide. Royce shared that he once worked, went home to a family, and owned a dog. When he lost his career, he simply gave up. The man sunk deeply into depression and isolation. He felt he couldn't go on and lost all hope.

Eventually, he lost his home because he couldn't afford it and couldn't find work. His depression fed his worst fears, and he became paralyzed by doubt and simply stopped trying. His wife left and took the kids and even his loving dog. His story almost sounded like a parody of a country music song. If I were not sitting next to him and seeing how lousy his life was, his story might have sounded like a bad joke.

Royce explained that he was sleeping behind a dumpster, and he felt an explosion of determination to get up one day. He jumped on a slow passing train. He thought that life couldn't get any worse and rode away from his despair. That decision to ride the train started his new life as a full-blown hobo. Royce didn't have anyone I could call. He didn't believe that anyone ever cared to see him again.

Royce quickly described the things he saw while living on a train car: caribou herds, lakes rarely seen by most people, and encounters with other hobos who were too often desperate and dangerous. I let him out at his spot to jump on his new train and gave him the bag of treats along with the remaining cash.

Royce thanked me for my help and took the bag but remarkably tried to return the money. He thought I left the cash accidentally. I told him it was for him, and he teared up. I don't share his story because I am proud of what I did; it was because of what he said as he was leaving:

"Some people are only one paycheck away from being just like me."

It was a chilling message.

After that encounter, I beefed up my efforts to save more, spend less, and be far more responsible with my income. I learned to be more grateful for the little I possessed, and while I was short on money for one week, I would do it all over again. This encounter is essential to so many of us because Royce was right. Most of us are one check away from disaster, and we should always prepare for the worst.

I have no idea how much longer Royce might have lived after that encounter. I never saw him again, but given his condition at the time, it is unlikely that he lived for long. While I hope that my assistance gave him enough help to live longer, a brief half-hour encounter certainly made a difference for that single day. This entanglement helped much more than I can say. I still cherish that encounter with Royce, even his awful smell.

I admit that we have to be cautious in our world, and some use their situations to access our sympathy; some simply

• CHARITY •

lie about their statuses. There is a measurable difference in the feeling when called to help others. Many years after Royce, I went to the aid of a man with a sign that read, "Hungry, anything will help." In this case, I bought food, water, and aspirin and added three dollars to the bag. I drove back to this man holding the sign, but he was only interested in the cash. When he saw the money, he turned and walked away without a word.

This encounter taught me a new lesson: not everyone is genuine when requesting assistance. I learned to measure the needs of those asking for help carefully, determine what is conceivably possible to assist them, and how much aid is achievable to give when I can help. In many cases, we want to help, but sometimes giving enables and worsens circumstances that we are not always aware of and assistance is not always appropriate.

In this case, Mardis might have been right about her thoughts on charity to certain strangers, but our world is complex. We should always be cautious and avoid becoming enablers of bad behavior. Certain behaviors, such as addiction, become a bottomless pit of need. I don't believe I'd stop again if I saw the same man with the sign.

When we assist others, we should never help to the point of being a doormat. We should also avoid manipulation by those who request it too often. A person in perpetual need sometimes needs to hear "No." We risk enabling bad choices and darker behavior that might not always be apparent and can sometimes have horrific consequences.

We can't solve problems by throwing money at them, but we can't do all of the work to solve problems we didn't create. While some people might need help, we can contribute many other ways. Resolutions only happen with effort, not by manipulating others to take the hit emotionally or financially. It might seem like a constant fight, but we learn from each battle.

Bad things happen that are not always our fault, but we have a choice to continue trying or give in to our despair. Royce admitted that he gave up, and that choice changed his life, but when our paths crossed, he taught me a valuable lesson that might help many more with their options. The desperation of Royce was like an unfed snake. It coiled around him, crushed him, and eventually consumed him. Sometimes the only way to avoid being crushed is to simply move.

To the apathetic, any charity seems like a handout to the lazy. Those who think this way are missing a critical point. Having the ability to help others or even inspire them implies that we are doing well. We are certainly better off than those in distress, so we should recognize how fortunate we are. Guiding and assisting another life to achieve more is an act of love and is one that we can share anytime. Some people will indeed manipulate others with their situations. Still, we have a choice, and with enough observation and time, we can make the right choice without feeling guilt and without demonizing others based on their appearance or needs.

Sometimes those people we wrongly believe have nothing to offer are the richest treasures of knowledge, motivate us to become better people, and strengthen us for the next battle. We should never worry about what we will get from each encounter, but when we share what we have with others in need, we often find that those in need help us more than we have helped them.

CHAPTER 9

Manipulation

I wish I could tell you that our emotional battles will ultimately end. The fight becomes more manageable, but we will have challenges as long as we live. There will be times when we will get so tired of fighting that we will surrender our better judgment and control of our lives to others. We do this because we want and need to trust that others are genuine and sincere, but not everyone will have respectful qualities or our best interests in mind. Along the way, many will take advantage of us if we give them the opportunity or if we surrender our will.

Anyone willing to manipulate others for personal gain has no decency. These people don't play by any rules; they will always bend them or completely break them. Their only desire is to gain an advantage, and they will use our trusting natures when it benefits them. Mardis, for instance, became a servant to an apathetic daughter.

• MARDIS •

Mardis suffered manipulation because she served a purpose, regardless of her declining health. When her usefulness ended, she was forgotten, used until her sad and tragic end. Years after she passed away, I got a call from an angry social worker about my mother, who gave my name as her new caregiver. I would never have agreed to such an undertaking, mainly because I traveled so frequently. The social worker told me that my mother accused me of abuse.

My mother and I were not in contact since before Mardis's funeral. During this time, she created a story that I was taking care of her, and she listed a long series of care responsibilities that I was neglectful of providing. Many of the things she listed were benefits she hoped to get from the state, including money.

After the funeral, our mother immediately admitted herself into an eldercare facility. She might have just been lonely or thought that she would continue to get around-the-clock care, although she was not physically disabled, beyond morbid obesity. Our mother contacted a social worker, but she was not honest about her living situation when she made the call. She attempted to gain similar advantages that other patients at her facility enjoyed.

I thought my sister was playing a prank when I first got the call. I even chuckled a little and asked, "OK, who put you up to this?" The social worker became angry and bluntly informed me that her call was no laughing matter. As it slowly sunk in that this woman was a real social worker and that she was serious about the allegations, I had to explain our dysfunctional relationship.

It had been years since I last saw her or spoken to her, and I was surprised she had my telephone number. Our mother didn't attend Mardis's funeral, and I was never a regular visitor when she was still living at her home because of our many disagreements. My sister was conferenced in the call and confirmed that I was not our mother's caregiver.

The social worker received the address of the facility where our mother lived and conducted interviews. Once she understood her residence in the facility, she realized that our mother was confused. I wasn't surprised that she would attempt to use her situation for personal gain, but trying to benefit without regard to the consequences and fallout was typical.

My mother didn't suspect that the social worker would contact me. I can't say if she created this story out of boredom, malice, or the attention she would gain. When confronted over this story, she told the social worker, "I had to do what I had to do." The social worker was gracious and attempted to help our mother find some of the benefits she wanted, but legitimately.

Mental health advocates confirm that diseases of the mind require a gentle approach to manage, but the collateral damage caused by the actions of the mentally disturbed can be hard to handle. Mental illness is not a fight that anyone should ever endure. Victims and bystanders trying to make sense of it all can never win, and it can be exhausting.

The methods and reasons for specific behavior are too complicated and perplexing to understand with any certainty. The unbalanced nature of mental illness is soul-crushing. It's terrible to witness a person having a breakdown, but the bystanders and targets of those breakdowns have to remain strong and endure a lot; it's not always easy. Although I could quickly forgive my mother due to her mental condition, I certainly would never be able to trust her or be willing to enable her behavior.

History reveals how seemingly ordinary people can make terrible choices, especially when they allow others to manipulate and influence their moral core dramatically. Dealing with a single person struggling with mental health issues can be particularly hard, but when a mentally ill person gains substantial power and influence, the consequences can be as extreme as genocide.

In the modern world, we see frequent examples of manipulation daily. There is usually a goal to control and a deep yearning for power. It happens in politics, religion, business, and everything in between. Historical examples such as the manipulation at Jonestown were awful, but one historical example from World War II caused a substantial loss of life and extreme cruelty: the Holocaust.

Shortly after World War II ended, society was still struggling with the reality that millions of innocent people were casually tortured, starved, and killed by soldiers at many death camps like Kulmhof and Auschwitz in German-occupied Poland. Entire families were forced into captivity, abused, and even fed to one another by the Schutzstaffel (SS), the Death's Head Units.

These Nazi soldiers held blind obedience to deranged authority figures, and they carried out thousands of murders per day. After the war, each of these war criminals stated only one defense at their military tribunals in Nuremberg: "I was just following orders." Nazi soldiers repeatedly claimed this while on trial for their crimes against humanity. Photographs and films were taken during the liberation of these death camps and helped to tell the tragic story of unimaginable suffering.

The evidence was incredibly striking and begged the question, were these soldiers psychopaths? How could anyone with a conscience allow such horrible suffering to happen or participate in it so casually? Most people firmly believe they would be utterly incapable of committing such atrocities in the modern world, but manipulation for evil is still possible. It happens right before our eyes and is well documented.

In 1963, Stan Milgram, a social psychologist, conducted many psychological experiments. One of his most famous was an obedience study. Milgram attempted to gauge the willingness of unsuspecting and average people to follow orders and commit horrible acts.

The obedience study held the potential to discover if the excuses given by Nazis were typical of sociopaths. Milgram attempted to understand how an authority figure could genuinely influence people to cause extreme levels of pain and suffering.

Yale University was the official setting, and Milgram and his staff were the authority figures. Ordinary people answered newspaper ads that offered $4.50 for help in an "educational experiment." In today's exchange rate, this was less than $40.

There was an implied sense of urgency because the study was limited. Getting paid was a motivational factor for many, and volunteers hoped to do a good job even though they knew minimal information about the task, an act of pride. Failure could also have been a factor in the manipulation process because the power of vanity is an essential tool for influencing the mind.

Milgram's first experiments involved only White men of various ages; many initially thought his experiments were biased. He later added additional ethnicities and women for a fuller range of results.

• MANIPULATION •

One of Milgram's lab assistants would play an essential part in the study by pretending to answer the same ad; the paid volunteer was oblivious, but the posing lab assistant was usually the last to walk through the door. When introductions were over, the next step in the manipulation process was to give the paid volunteer a "random" choice of roles to play, teacher or student.

An official lab technician would hold written roles to play in his hands, and in every case, the paid volunteers would choose first and always play the teaching role.

The paid volunteer, now teacher, had one job: Ask questions and flip switches for incorrect responses as a consequence. When activated, the bulky controls would allegedly deliver shocks to the student for each wrong answer. The voltage ranged from 15 volts to 450 volts, which meant that each incorrect response would provide a higher voltage and cause more pain.

Every paid volunteer was given a brief electrical shock at the lowest setting for context and fairness. While the shock was uncomfortable, the volunteers were fully aware that the role of the teacher was to harm. The lab assistant, pretending to be the student, sat in a nearby room outside of the paid volunteer's view as the experiment began.

The lab assistant pretended to do well for a few questions, but he used a hidden tape recorder with prerecorded complaints of pain for failed replies. The recordings were screams, gasps, warnings of heart trouble, protests, and demands to be let out of the room. This same taped series was used with many volunteers answering the ad that endured the same manipulation in this experiment.

In every experiment, volunteers believed that they were causing harm to the student for their failures and tried in subtle ways to help. Yet, the volunteers continued to flip each switch, clearly marked with higher voltage and potentially inflicting more damage to someone they had only met moments ago.

The lethality of voltage to the human heart ranges from 100-250 volts. Still, volunteers continued flipping switches to 450 volts even when the student no longer replied, simply because the authority technician with the clipboard told them to continue.

During the experiment, when any volunteer showed concern, signs of fatigue, or any apprehensions about causing more pain, Milgram's staff would calmly say, "Please continue the experiment." The majority of the volunteers continued when asked to do so; they were just following orders. A staggering sixty-five percent of those who participated in the experiment were willing to administer what they believed were shocks, ranging from slight discomfort to death.

When asked why they continued, they would say, "That guy told me to keep going!" Milgram showed that most people were willing to take action that would cause pain, and they fought through their apprehension to do horrible things when they blindly surrendered their will to an authority figure.

In terms of obedience, Milgram suggests that humans follow two mental states. The first is the "autonomous state," where people will take responsibility for their actions, speak up, or flatly refuse to continue causing harm. The other is the "agentic state," where a person will listen to an authority figure, do their bidding, and defer any consequences for their actions to the authority that gave them the order, regardless of how terrible, even when it meant potential death.

This study is significant because agentic behavior continues in our world today, and there is no shortage of those willing to use that behavior for personal gain. Misleading information, hoaxes, and conspiracies nurture an agentic frame of mind and happen under our noses.

Computer hackers and foreign governments continue to use these tactics to gather personal information to access networks, drain bank accounts, cause social unrest, and manipulate behavior and belief. The methods employ an authority figure, a sense of urgency, an offer of something beneficial, and the threat of consequences for a failure to act. When we turn over an immediate trust and fall for one or more of these methods, we grant access to our minds and computer networks. An authority commands that we take action, and we do, turning over our control.

Today we mostly use the term "hacked" to describe our cooperation. It happens every day, thousands of times per day, because we fail to think in a critical way when criminals manipulate us to gain our cooperation. Authority cybercriminals pretend to be someone of importance, and our relationship and trust in our technology do the rest. Making a single miscalculation and acting in this agentic state today has grave consequences due to our connected world's nature.

Some unscrupulous people have even built a business on the art of manipulation, which is worth billions of dollars a year. The technology that we can no longer live without is now the most efficient tool for that purpose. Technology is a synthetic form of entanglement. We can use technology with good intentions, but for evil just as quickly.

Profiles are created through our social media by asking basic questions. We freely reveal our critical relationships, names of family members, telephone numbers, and other personal information to access those social tools. We trick ourselves when we use fun and seemingly harmless online quizzes. We are willing to turn over information about ourselves that creates psychological profiles with each answer. These profiles help refine algorithms for further and deeper levels of manipulation. Cybercriminals are often successful, but it all starts with our impatient nature.

Most people are under the impression that they are doing something harmless or a fun activity. Still, the information gathered targets our buying habits and political beliefs and creates a false consensus for various purposes. Ultimately our behaviors are influenced by trends.

Digital manipulation in the form of a false consensus happens when we falsely believe and trust online opinions because of the volume of replies. The majority of these responses are, preprogrammed, artificial intelligence, and bots, not living and breathing people in most cases. These programs can also create conversations based on keywords and replies, and many users believe this sharing is genuine.

Using an avatar and written commentary is usually the key to acceptance. Humans are often too rushed to vet users we interact with online thoroughly. Even reviews and polls are manipulated and encourage a bandwagon effect to sway public opinion.

This type of planted information usually requires vast resources. With enough financial means, organizations can employ biased data, and they can quickly have an enormous impact on societal beliefs that suit their needs. Scheming manipulators use technology and psychology to gain an online social advantage by creating those trends. When their plans are in motion, the erroneous data rapidly gives organizations and others the power of persuasion and can be powerful enough to destabilize world governments.

The mechanisms once used to protect against this manipulation were dismantled, and the erosion of security and truthful information is only worsening. Ambiguous fearful concepts, used with consistent buzzwords, become talking points. Anytime these talking points are said enough, an ideology becomes believable to many, and ordinary people will act in an agentic manner. Once adopted by regular people, ordinary citizens begin to parrot back this same disinformation as if the learned talking points are original thoughts.

A policy from 1949, the fairness doctrine of the US Federal Communications Commission (FCC), was once used to protect information consumers. After World War II, potential manipulation and influence by foreign governments was a significant concern for national security. The fairness doctrine policy required broadcasters to present honesty in broadcasting or face fines and punishments, such as removal from distribution. During the 1950s, a series of threats between the US and the former Soviet Union made this policy the utmost importance in the United States.

There was also censorship of specific material by ethics committees, monitoring broadcasts for obscene programming while protecting freedom of speech. Finding no other way around these fairness standards, manipulators dismantled those safeguards in the 1980s. The lack of safeguards created a new environment where abuse of speech, irresponsible, and untrue information could gain an unfair advantage and thrive, legally and without consequences.

As the fairness doctrine policy ended, radio broadcasts of sponsored opinions and extreme ideologies became frequent.

• MANIPULATION •

Opinion broadcasts were slowly characterized as news in print, radio, television, and eventually the internet. The internet and cellular systems were the final steps in creating an environment where organized groups could quickly and efficiently manipulate a broader consumer audience. Much of this work was born out of experiments like those done by Stan Milgram.

However, a modest percentage of people are still unwilling to act in an agentic state. Manipulators can only entice, but it is up to the individual to avoid negative influences and a mob mentality. The mob has difficulty understanding an autonomous personality. An agentic mind frequently fails to act with caution or self-control, and those that feel empowered to cause harm will. When conflicting ideas for civility clash, the agentic mind feels attacked and becomes brazen, reckless, and dangerous.

In 2021, a group of men and women stormed the US Capitol Building in Washington, DC. Provocateurs encouraged men and women using a false consensus to nurture a belief that personal rights were at risk or stolen. This drip of disinformation took months, possibly years, while stoking anger. Ordinary people were manipulated and became pawns for the sake of power for a select few.

Ultimately those acting in an agentic state during this manipulation campaign saw their lives significantly disrupted and, in some cases, ruined. They failed to consider the personal consequences of their actions but turned over control of their better judgment to serve the needs of an unworthy few. The primary defense of each person's arrest was following orders, being caught up in the moment, and arriving by invitation to show their patriotism.

To think critically and independently about any situation and avoid being manipulated, we must ask many detailed questions, observe the consistency in each answer, and, of course, have clear evidence before accepting claims. It takes time, and it is against modern behavior. Although many live through blind faith and follow ideologies they don't fully understand, consequences still exist, and no one should be surprised when those consequences surface. While some of the impacts seem narrow, those consequences will reach and impact many generations over time.

Many honest people still exist and are willing to give sound advice. Reliable sources of information might feel harder to locate. Only now are some beginning to realize that the massive accumulation of knowledge doesn't mean that the quality is helpful or always benevolent.

While entanglements can bridge natural connections to one another, technologies used today attempt to force those relationships and use manipulation techniques. Those seeking control will always attempt to build false beliefs if it serves them, but they never seek to achieve balance or help others.

Earlier communication tools, when used responsibly, might have amplified feelings and emotions in productive and

• MANIPULATION •

positive ways. Every person who contributed to a broadcast was once part of something incredible. A writer might have felt emotionally inspired to write, an actor might have been motivated to perform, and those responsible for content airing decisions were part of an intense entanglement but with safeguards. Each contributor shared their vibrations through their art and took responsibility for delivering it with seriousness and pride; consequences prevented recklessness.

Manipulation isn't always blatant, but our pride and vanity make us think we can handle any outcome. We convince ourselves that we are invincible and can do anything, but we are usually in such a hurry for satisfaction that we don't reflect enough on the long-term impact of what we do in the short term. Manipulators realize this too and will use that to their advantage.

There are always unfulfilled promises, threats against inaction, conditions, and other psychological tricks filling our minds with anxiety and the urgent need to act immediately. Still, we must not allow others to argue with us to the point that we give in. Sometimes, the answer that we give should simply be "No." Responding this way and taking more time to consider what was said is better than merely reacting. Even though making ethical and informed decisions takes longer, it is vital to our long-term survival and growth.

CHAPTER 10

Restoration

I feel many alignments and connections exist beyond our present understanding. If I look back 500 years, I can't possibly know who my ancestors were, what they were like; what they enjoyed doing, or the decisions they made, which ultimately led to my existence today. If I look forward 500 years, I can't know who my descendants will be or how they might behave either. Yet, those "who were" and those that "will be" are connected to me today.

Each day, we walk the road of time with blinders and have a narrow view of one another and the world in general. When we look beyond and outside of our small circle of friends and family, we can see billions of other people, and we share this world with each of them. We will never know the vast majority of those souls during our time on this planet, but we feel their impact on the world in big and small ways. Our presence and the decisions that we make will also impact the descendants of those unknown billions too.

• MARDIS •

Each of us connects during our present-day through our every decision. Our actions and others help establish relationships in the present and lead us to new paths of existence, even simple actions by those we never meet influence our lives. As we make our way along the trail of time, we may find both opportunity and failure, but change will always be inevitable, so we should do our best to make every change count for something worthwhile and beneficial.

We have so much potential to extend ourselves beyond our limited understanding. The universe is certainly large enough. There are an infinite series of possible outcomes and versions of who we might become with each decision we make.

All lifeforms can share, heal, and connect with us when our spirits reach out. As the essence of life attempts to lure us to a new way forward, it is not until we stop and look at that new road ahead that the doorway to the past closes behind us, and we have no choice but to continue on each new path.

Our universe changes with every life, death, and loss; we realize the vibrations and tones that contribute to the soundtrack of our present reality are different. Our future potentials also alter, even as we continue to offer hope and possibility.

We might find luck and lose it many times, but when we dedicate life to negativity, we plunge ourselves into a dark existence. Yet, even in the depths of that darkness, we can still be reached and set free. We simply have to reject conforming to the patterns of darkness and hopelessness. When we do that, we transform ourselves. We are never too far gone, and escaping the dark is always possible.

Turning back time is not an option. Our species only has one way to move, and that is forward. As much as we would like, we can't physically go back to a point in time to correct our mistakes. Attempting this would require us to recreate the precise conditions and vibration for all things in the universe, just as they were for a single moment. Although it might sound like a good idea to go back, it is far more work than it is worth and filled with extreme uncertainty.

The grandfather paradox is a thought experiment where we ask: if we travel back in time to meet our grandfather before he met our grandmother and stop the two from marriage, how can we be born to take the journey in the first place? The answer fits somewhere in the principles of uncertainty, a significant part of each quantum reality.

Some believe if we tried to stop that union, we might cease to exist and just fade away. However, our quantum reality is playing a slightly different vibrational tone. We would remain unchanged in our original departure gate universe. In other words, we'll continue to exist, but we would prevent what could have happened in another parallel universe.

Even if we somehow could escape the bonds of time, we can't escape the uncertainty of each of the infinite realities created by billions of living people during that period. We would have no way of knowing where we might end up. A specific place may look similar, but tiny variations would exist. Every variation is a clue and indication that we went to the "wrong address." Traveling back a few seconds would require more calculation than might ever be possible in a physical form and can't be done quickly.

Finding and knowing the precise quantum vibrational pattern for a single moment is more complicated than capturing a dust particle with tweezers on a beach in a Category 5 hurricane. So our only real and practical hope is to continue moving forward with acceptance and determination for a better future while making better choices.

A happier life requires understanding our past and accepting our failures. We need to acknowledge our present, whether our circumstances are good or bad. Acceptance will always be essential in initiating any change in our existence. We can't change what we hide from and avoid. We should also consider what our existence means to the future and what we will leave behind. We shouldn't leave the world a mess because we were too lazy or confused to try to do better.

No one knows for sure what will happen, but the best guidance has been: "just shut up and calculate," I do believe that it might be possible that our spirits can communicate with or replay precise moments in time. In the flesh, the task of traveling through time would require many years of technological advancements, but in the spirit, much more might be possible. If we follow the energy conservation laws, we know that energy can't be destroyed, but it can redirect. There might be some redirection if we think of our spirit as energy.

Suppose we can communicate on the quantum level and spiritually match a past vibration. A specific momentary match with accuracy for any individual might be plausible. In that case, it might be possible to replay them like a recording. However, the instant the two moments in time connect, they would amplify quickly, alter the connection, and entropy would terminate that relationship. Playing back a moment might be too quick to understand, reason with, or interact with in meaningful ways.

Quantum-styled recordings could also be one of the many reasons some people see ghosts. Typically, spirits are said to perform particular tasks on a kind of loop, like a ghost rocking in a chair or other repetitive behavior. Certain people might behave as the necessary "hardware" to connect and replay historic life events. They may be capable of projecting smells, sounds, and visions captured in time, which many consider restless spirits rather than recorded moments from energy in time.

These moments would always be short-lived because once the realization hits that something is happening and the frequency changes, those unique connections would be out of sync and collapse or out of phase moments later. When you look for a specific moment, you can no longer measure what was there.

A similar phase transition happened to me when I was in elementary school. I was in my school talent show; I was in the fourth grade. My mother dropped me off at school that night and instructed me to wait outside after the show. She said she would pick me up later.

When the show was over, the school's parking lot had a steady flow of families departing for home. As the last family was leaving, they asked if I would be OK. I assured them that my mother was on her way. I waited alone in the darkness for a little more than an hour and a half, and just as I began to feel stressed and uncomfortably nervous, I heard the sound of a man's voice saying: "Your mother will be here in a minute!"

There was no one around. The school was completely closed off, there were no cars, and there was only one way to enter. The school was on a dead-end road. I nervously yelled back, "Hello?" But there was no answer. The entire area was dark, and the only lights were the stars in the sky. The man's voice sounded like a moving echo reverberating off the road in what sounded like a passing car.

• RESTORATION •

After a minute, I looked down the long stretch of darkness, and I saw headlights moving in my direction, and the phantom Samaritan's assurance was correct. It was my mother flashing her headlights to let me know it was her, finally arriving to pick me up.

I was driving by that same elementary school more than twenty-five years later. The area had many improvements, and the road was no longer a dead-end. Just by chance, I accidentally turned onto the new portion of the road. I was traveling with friends and told them about that night and the weird sound of the man's voice. They began to tease me.

My friends considered it a ghost and asked if I heard banjos playing because the area was still rural. Although the area now possessed many improvements, it was still in the middle of nowhere. As we drove closer, I pointed to where I was standing on the night when I heard the mysterious voice. I yelled out of my window to demonstrate what was said: "Your mother will be here in a minute!" And it hit me like a bolt of lightning: the voice I heard that night could have been my own.

I can't recreate that incident, and I am honestly not sure how it happened. Did I somehow reach out across time to give myself a message? In all sincerity, I just don't know. One friend suggested that I heard the voice of a ghost, and the other considered the voice an angel. Others might view this event as déjà vu, but all I can say for sure is that the memory stayed with me for many years, and when I demonstrated to others as an adult, it felt as though the two points in time connected in a kind of spark.

Strange events like these usually came with an odd yet mild visual blacking out and a kind of dizzy feeling, so it could be that my brain was suffering from an undiagnosed abnormality. The simplest explanation was that it was all in my mind.

Although moments like these also felt spiritual and happened frequently, I never put them together until the passing of Mardis. As a joke, I would usually ask, "OK, what did I do the last time?" Each time I attempted to measure the déjà vu that moment collapsed. What happens next might no longer be possible. The simple act of observation and acknowledgment changes those circumstances. A new reality might have been born and caused destabilization to those points in time.

One might argue that I recalled past life experiences, false memories, or many other phenomena, including entanglement. The only way to know more about these moments would be by testing the many assumptions, which is tough to do because they never seemed to happen on a schedule.

One of the most powerful moments occurred a month before my thirteenth birthday, and my estranged father came back into my life in a similar unusual manner. During that time, I wished that he would return. I hoped he would save my sister and me from the hell we endured for so long, but he never came to our rescue. However, before my birthday, I felt a strong determination to see him again. He was on my mind every day, sometimes hourly.

I prayed, searched through old phone books at libraries, called long-distance operators, and wrote letters to a family member who might have known his whereabouts, but my father continued to live his life on the road, and few knew how to find him. Most people I spoke with said that he would usually just show up out of the blue. Finding a specific address for my father or even connecting with him by phone was highly remote.

Still, my focus was heavily on seeing him, and my spirit wanted to know why he never returned. I felt this weight in every cell of my body. A visit from my father was all I wanted for that birthday. I needed to share the hell we lived through because he was absent. I called out with every ounce of my faith, continuously meditating for days and focusing on just one more conversation with my father. I didn't know at that time if he was alive or dead.

Although I loved my father, seeing him was more about closure. I wondered what he might have said about our lives or the abuse that we endured. I wasn't angry, but I was determined to understand if we were burdens or worthy of saving in his eyes. I was desperate to hear from him to get answers, but he would know that we never forgot him.

Five days before my birthday, our father snuck back into town with a new wife and a baby girl. I couldn't believe it when I saw him; it was like a dream. He explained that he was driving through Texas but urgently felt that he needed to go back to Florida. When he did, he was only in town for a few hours before he found my sister.

My sister helped to coordinate a secret visit with him and me. Our mother eventually learned that he was in town. She surprisingly offered our dad a job as a maintenance man for her boyfriend's business, and the job included a place to live. We knew that our mother wanted to present herself as a grand success, but she never shared the cost of her success, which was turning a blind eye to her perverted boyfriend sexually abusing her children or her brush with the law that followed.

While offering a home and employment might sound noble, our mother's offer of assistance meant my father would have to jump many hurdles. He managed to handle the stressors of life for a little more than a month before he felt his wanderlust call him, and he hit the road again. Still, there was one final Thanksgiving with him, even Mardis was there, but he just couldn't handle the stress of staying in one place for too long or the constant need to grovel.

• RESTORATION •

A few weeks later, I was able to spend Christmas with him in an abandoned house. Then, the following Easter, his life came to an end. He made his final journey alone, hitchhiking to the Veterans Affairs Medical Center a hundred or so miles from his most recent bed. My mother paid for a one-way flight for me to see him, but deep down, I felt that it was too late. I traveled alone, barely thirteen, and I didn't know how to handle this revelation.

I managed to get to the VA Medical Center in Jackson, Mississippi. The solemn trip was my first flight, first cab ride, and first time alone in the world away from Mardis, my mother, and my sister. At this VA Medical Center, a nurse seemed to know who I was and picked me out of a crowded waiting area. She told me the floor number, and as I stepped off the elevator, my stepmother was crying in the hall. It all happened so fast. I didn't know which room he was in, but I instinctively walked straight to it.

A doctor walked in behind me and slowly pulled back the sheet; it was dad. My father was only recognizable by his perpetual five o'clock shadow and a Playboy bunny tattoo on his leg. His appearance was not what I hoped to see. It was not a peaceful look of rest. My dad was always skinny, but he was bloated and unrecognizable. He had tears of blood coming out of his eyes, and his tongue was sticking out of his mouth like he died choking and gasping.

I didn't want to believe that this was the same man who returned with that strong feeling to find us just a few months earlier. How could he leave so quickly, only to die and look this way? I was and still am heartbroken by his death. I later learned at his funeral that he tried to get help, but in the end, he sacrificed his ability to see us. He knew that we would have a much more stable life with Mardis than living in a van, on boxes, or in tents.

I also learned that he knew Mardis well. Although she never spoke of him, Mardis played a role in bringing our father and mother together. He sincerely believed that Mardis was our only hope for survival. Years after his death, my sister surprised me when she confessed that she began to see our father at various places after he passed away, and she felt like he was watching over her. She kept it quiet for many years. I was surprised at the time, but now I am a little more generous in my belief about such matters.

Although I didn't have a final goodbye in spiritual form from our father, I at least encountered one last valuable living moment with him, and I am grateful for that opportunity. I would rather have my father alive, but sometimes, no matter how much faith we have, the answer is simply no.

Things might have been different in another universe, but I continue to pursue hope in this one. Today, I feel the connections that we all share. Parents, teachers, a bus driver,

• RESTORATION •

and a collection of friendships have restored my faith and trust in so many things, and for that, I will always be grateful.

I once felt like I was alone, but I know now that I never was. When I was down, the people who came to me always knew what I needed to hear and what I needed to see; they guided me. I just needed to allow those people into my life and put in the effort required to grow.

These days my life is filled with body scans, blood tests, and many more distractions, but I feel renewed. While I am here, I'll make every minute count for something good, and I'll answer any call for help I manage to hear. Time grants us a limited amount of life; we should never squander it.

As time passes, we recognize how valuable every moment has been in our lives. Even the many disagreements with Mardis were helpful in one small way or another. She helped shape me to become a better person. Forgiving others only seems complicated, but when the weight lifts, it unleashes our spiritual potential, and we become forces of nature. Once we realize that everything has a purpose and helps us evolve, we'll solve many problems and pave the way for others to find their paths to peace. For now, the best I can do with the time I have remaining is to carry on.

If You Need Help

This book has revealed my journey of survival. No person, child or adult, should deal with any kind of abuse—physical, mental, emotional, or other. If you or someone you know is enduring abuse, please contact your local officials. If you or someone you know is addicted or has mental health issues, please contact a professional for help.

Abuse

– Call the local or state child protective services.
– Call the National Domestic Violence Hotline at 1-800-799-SAFE (7233). Online, go to thehotline.org. The Family Violence Prevention and Services Act operates a 24-hour, national, free hotline. It provides information and assistance to adult and youth victims of family violence, domestic violence, or dating violence, family and household members, and other persons such as domestic violence advocates, government officials, law enforcement agencies and the general public.

Addiction or Mental Health

– Call the local or state alcohol and drug addiction hotlines.
– Call the SAMHSA national helpline at 1-800-662-HELP (4357). Online, go to samhsa.gov. The Substance Abuse and Mental Health Services Administration is a free, confidential, 24/7, 365-day-a-year treatment referral and information service.

CPSIA information can be obtained
at www.ICGtesting.com
Printed in the USA
LVHW081620020422
714873LV00005B/96